BUILDING SCOTLAND

BUILDING SCOTLAND

Celebrating Scotland's Traditional Building Materials

Edited by Moses Jenkins

Foreword by Rt Hon Alex Salmond MP MSP, First Minister of Scotland

JOHN DONALD

in association with Historic Scotland

First published in Great Britain in 2010 by
John Donald, an imprint of Birlinn Ltd

West Newington House
10 Newington Road
Edinburgh EH9 1QS

www.birlinn.co.uk

ISBN: 978 0 85976 710 1

This publication is available from
Publications Department, Technical Conservation Group, Historic Scotland,
Longmore House, Salisbury Place, Edinburgh EH9 1SH
Tel: +44 (0)131 6688638, Fax: +44 (0)131 6688669

Commissioned by Technical Conservation Group

British Library Cataloguing-in-Publication Data
A catalogue record for this book is available on request from the British Library

Typeset by The House, Edinburgh
Printed and bound in Britain by Bell and Bain Ltd, Glasgow

CONTENTS

FOREWORD

When we think of traditional buildings in Scotland, our mind often leaps to images of castles, palaces and stately homes. But traditional buildings are more than just these high-status buildings; they are also the cottages, farmhouses and terraced tenements that make up our villages, towns and cities. They are the buildings that form part of the world around us, that create our sense of place, and are a pivotal point of reference for understanding who we are today.

The rich and appealing variety of styles and materials in our towns and villages across the country has evolved over centuries due to variations in the local materials that were available and building traditions handed down from generation to generation. Our traditional buildings were built of local materials by skilled craftspeople with generations of experience behind them to work in harmony with the environment around them. The traditional buildings that survive today are testament to the quality, durability and versatility of Scotland's natural resources and the skill of the generations of craftspeople who not only built these structures with their bare hands, but who also won the raw materials from the ground and whose contribution to the buildings around us is rarely recognised.

Our traditional buildings tell us much about where we have come from. Scotland's traditional buildings, and the materials used to build them, are a reflection of our culture, our society and our economy. But more than that, they are a reflection of our success as a nation – of our talent, our practicality, creativity and ingenuity.

Rarely do the generations of craftspeople who built so much of the world around us and who won the raw materials from the ground receive the recognition they deserve. It is to these people, and the people who keep these important skills alive today, that this book is dedicated.

Rt Hon Alex Salmond MP MSP,
First Minister of Scotland

ACKNOWLEDGEMENTS

Historic Scotland thank the following for their help in the production of this publication: NERC BGS, The Scottish Crannog Centre, The Scottish Lime Centre Trust, The Scottish Reformation Society, The National Galleries of Scotland, The Royal Commission on the Ancient and Historical Monuments of Scotland (RCAHMS), The National Museums of Scotland, Charles Laing and Son, and The Great Steward of Scotland Dumfries House Charitable Trust for permission to take photographs on their property or to use images for which they hold copyright. Thanks also to Mairi Sutherland and all at Birlinn publishers for their hard work in preparing the book for publication. Thanks too for all the help received from many colleagues in Historic Scotland. Lastly, thanks to the authors and those who assisted them in the preparation and editing of the individual chapters.

This book is dedicated to Noah Jenkins, born October 2009.

INTRODUCTION

Much has been written on the subject of the historic architecture of Scotland. In contrast, there has been relatively little written about the materials from which these traditional buildings are constructed. It is this deficiency in both knowledge and appreciation of the basic materials which have been used to construct the buildings in which we live our daily lives that this volume seeks to redress.

This volume is intended to be a celebration of fourteen of the most significant traditional building materials of Scotland. Many, such as stone and timber, have been used since the earliest human inhabitants came to Scotland. Others, such as pantiles and iron, have come to be used in more recent times. For the majority of our traditionally constructed buildings it is the use of these materials in combination with one another that has produced such beautiful and diverse results: stone, lime and pantiles come together to form the distinctive buildings in our East Neuk towns; timber, clay and thatch combine in other areas with glass and iron used together in some of our most impressive monuments.

The book is presented in fourteen chapters, each written on an individual building material by an expert in their field. As would be expected in a book with such a diverse range of authors, each one brings their own unique writing style to the material they discuss. Despite these subtle differences there are clear themes in all chapters in terms of the subject areas covered. For each material, consideration is given to the history and variety of its use, how the material is manufactured or won and regional variations.

Richly illustrated with over 100 specially commissioned colour photographs taken throughout the country, *Building Scotland* not only aims to be informative but to visually capture the colours, textures and spectacular uses of Scotland's building materials. It is hoped that this volume will allow a greater appreciation of the materials which have literally been the building blocks of our rich built heritage.

Moses Jenkins,
Editor

CHAPTER 1

TIMBER

Geoffrey Stell

. . . shaping the nature and development of Scottish building traditions . . .

In Scotland, as in the rest of the temperate world, timber has formed a significant, perhaps even a central, part in shaping the nature and development of the country's building traditions. Until comparatively recently, however, the range, variety and chronological depth of timber construction in Scotland remained relatively understated and to some extent underappreciated, even among those with specialist interests in Scottish building practices. Happily, all that has now changed, and more than two decades of intense environmental and ecological awareness have seen a considerable upsurge in the study and scientific analysis of woodlands and their exploitation, and of the many and varied historical uses of timber in Scotland. Concurrently, there has also been a dramatic increase in the application of home-grown timber, not just for general building purposes but also for many showcase modern architectural designs, often as external cladding. At the beginning of the twenty-first century, wood is once again a building material of the present and the future, not just of Scotland's past.

Although limited in its external use over the past two or three centuries, timber was historically one of the most ubiquitous of all building materials. Not only has it been the principal structural component of flooring, internal framing and roofing, it was extensively employed for plenishings, such as panelling and other forms of wall lining, and for substantial fixed or semi-fixed furnishings, for example, benches and pews in large public buildings such as churches and courtrooms. Until recent times timber has been equally in demand as a key aid to construction,

access and transport on virtually every building site. Most architectural and engineering projects of any size would have consumed prodigious numbers of poles and planks for scaffolding, walkways, arch-centering and shuttering, as well as for pallets, carts and sleds; even brushwood regularly played its part in, for example, the creation of access tracks. Historically, the use of timber in building has been almost limitless in scope, while the special constructional, industrial and military needs of wartime Britain in 1914–18 and 1939–45 intensified such use at the very time when submarine warfare rendered its importation from abroad increasingly difficult. The result, especially in World War II, was an intensive and extensive exploitation of native woodlands almost to the very limits of Scotland's resources.

Since 1945, considerable advances have been made in techniques of archaeological detection, particularly in the identification of relic post-holes and beams, and in increasingly sophisticated scientific analyses of timbers and their carbonised residues through tree-ring and radiocarbon dating respectively. One result has been that during the past half-century our academic knowledge of the structural use of timber in the prehistoric and medieval periods in Scotland has been dramatically enhanced. The crude adaptation of natural wood for structural purposes may conceivably go back to the very earliest tent-like habitations of the hunter–gatherers of the Mesolithic era but, thanks to archaeology, we can now confidently assert that roofed buildings of some magnitude that employed wood converted to timber have a much more ancient ancestry in prehistoric Scotland than had previously been thought possible.

Trees have been utilised
for building purpose since
earliest times

Excavations carried out between 1977 and 1981 at a crop mark site at Balbridie on Deeside, Aberdeenshire, provided a first set of key landmark findings. Here, the excavators revealed convincing evidence of a storage barn or granary that evidently employed squared timbers to create a substantial rectangular aisled structure of a scale, style and date hitherto, and still, unparalleled among the Neolithic building cultures of the British Isles. Measuring externally 24.6m in length by 13.4m in width and dating from the early to mid-fourth century BC, the Balbridie barn was a remarkable timber structure by any standards. In size, proportion and structural detail, including its bowed end-walls, it bore a resemblance to the earlier of two timber halls excavated in the mid-1960s at Doon Hill near Dunbar, East Lothian. This equally remarkable structure was ascribed to an early seventh-century, pre-Anglian phase but, even at that early 'Dark Age' date, it was merely heir to a timber building tradition that went back several millennia.

Various categories of late prehistoric and early historic structures form part of that emergent tradition. Even stone-built roundhouses are now understood to have made significant use of timber, especially for roofing and framing purposes. Insular sites known as crannógs or loch dwellings, which are particularly numerous in Ireland and Scotland, appear to have been even more prodigal in their use of timber for piles, gangways, house platforms and palisades. The very name given to this form of settlement, perhaps derived from the Old Irish 'crann' (tree), may itself faithfully reflect their wooden character, which has certainly been abundantly attested by excavations in Loch Tay, and can now be clearly appreciated in the careful and authentic reconstruction of a crannóg of this type at Oakbank, Perthshire.

Crannogs were early examples of what could be achieved using timber engineering
© SCOTTISH CRANNOG CENTRE

The simple uses to which timber could be put in the form of tied poles

Timber-laced internal reinforcement and external facing of drystone or earthen walls was indeed a characteristic structural technique of much late prehistoric hill-fort architecture across Europe. In Scotland, such intra-mural frameworks comprising horizontally laid layers of intersecting longitudinal and lateral wooden beams which projected through the wall-faces were first clearly revealed in late nineteenth-century excavations at Castle Law, Abernethy, whence it came to be known as the 'Abernethy' tradition. Such a form of construction was similar to the stone-faced defences with exposed timber beam-ends that were observed and described by Julius Caesar in the course of his campaigns in Gaul in the mid-first century BC, and to which he applied the term *murus gallicus*. Many such structures were on a massive scale and it is now known that, to effect greater stability, some made use of jointed timberwork and metal spikes or nails, techniques which are known to have been independently reinvented and applied in early Scotland at the late and large forts of Dundurn, Perthshire, and Burghead, Moray.

A number of timber-laced hill forts in Scotland and France also provide the main concentrations of the related phenomenon of vitrification: that is, the molten fusion or shattering of stone resulting from intense, volcanic-like heat of around 1,000°C. Ever

since the late eighteenth century, when vitrification was first recognised on sites such as Craig Phadrig, Inverness, and the massive and lofty Tap o' Noth, Aberdeenshire, debate has raged over the possible circumstances under which such conflagrations might have taken place, and why the forts were vitrified. Essentially, the arguments have ranged between whether the process was a deliberately defensive creation on the part of the occupants, in order to present an impenetrable wall of molten material, or whether it was a deliberately punitive act carried out by triumphant aggressors intent on wantonly transparent destruction. What has never been in doubt is that timber, almost certainly in prodigious quantities, would have been an essential ingredient in the burning process, and would have been a major structural component of the walls of the forts before they became vitrified.

Structurally and symbolically, timber on its own, or in combination with dry or mortared stone, continued to be a building material of the first importance as Scotland moved, mainly from the reign of King David I (1124–53), into the international arena of medieval church and castle building. As elsewhere in Western Europe, the first phases of 'feudal' settlement in Scotland witnessed the construction of numerous motte or motte-and-bailey castles in and around which the main structures,

Timber has been used to form structures both prestigious and humble. The latter is represented here in this cruck frame, once a common building form in Scotland.

including towers and palisade-type defences, were almost invariably timber-built. Such sites now survive mainly as earthwork mounds and ditches, but archaeological techniques that have become capable of detecting subtle evidence of post-holes and beam-slots or trenches, have clearly revealed the palimpsest of a once extensive timber-building culture that has disappeared from view. Indeed, among certain categories of structure, such as halls and hunting lodges, there are indications that, for the Scottish nobility, timber long remained a material of status and preferential choice, not one merely of convenience, ease and speed of construction.

The one-time timber profile is also indicated by innumerable roofless and floorless ruins, where timber has not been confined to buildings of subordinate, ancillary purpose. The lengths and spans of many stone-cased buildings, such as Morton Castle, itself probably a sophisticated fourteenth-century hunting lodge, also clearly reflect the use of substantial timbers, in this case from Nithsdale, a valley whose clay soils have long been an important source of oak trees. The various raggles and scars in the surviving enclosure walls of mighty castles, such as the late thirteenth-century Comyn stronghold of Inverlochy, show quite clearly that timbers, often huge scantlings, were used in the long ranges that ringed the internal courtyard, testament to timber building on a truly impressive scale, and which gives the lie to their modern appellation as 'stone' castles.

However, medieval timber building in Scotland is by no means a world of lost materials and vanished traditions. Among the well-wooded landscapes of deciduous trees depicted on the late sixteenth-century maps of Timothy Pont is the area around Darnaway Castle, an early royal hunting lodge associated later with the earls of Moray, which stands to the west of Forres, Moray. Darnaway's sessile oaks were clearly of an eminently usable quality and, importantly,

OPPOSITE: The impressive hammer-beam ceiling of Stirling Castle's Great Hall, recreated in Scottish oak

were accessible via the nearby River Findhorn, thus allowing them to be transported by water to the coast and thence to the south. A significant source of timber over many centuries, particularly for a number of major royal building ventures during the reign of King James IV (1488–1513), it is entirely appropriate that Darnaway Castle retains the earliest known and most complete medieval timber roof in Scotland.

In a scheme completed in 1812, the medieval castle of Darnaway was turned into a Georgian mansion, but the transformation carefully retained and deliberately 'showcased' the medieval great hall and especially its magnificent roof which, even at that date, was regarded as a precious object of venerable antiquity. Not until 1987, however, when it became accessible for close-range survey and dendrochronological analysis, could its actual antiquity be established. Tests on a suitable set of twelve oak timbers pointed conclusively to a felling date of around 1387, so, given that oak must be worked whilst it is green and before it hardens, the roof members would thus have been assembled and placed in position shortly after that date. With an overall length of almost 27m and a clear span of nearly 11m, Darnaway is close in general scale and style to the roofs over the great halls at Stirling and Edinburgh castles but, as we now know, also in part thanks to dendrochronology, these royal roofs were not erected until the early sixteenth century, more than four generations later than the precocious Darnaway. Even in England in 1387, the hammer-beam genre, an alternative means of spanning halls that would otherwise have been aisled, had still a few years to run before reaching its apogee in the reconstructed Westminster Hall roof.

As it now survives, the Darnaway roof is essentially sui generis, a two-tier structure with an independently constructed double-collared upper stage. The lower stage is of truncated arch-braced collar beam form but lacks vertical hammer posts,

hence its classification as a 'false' hammer-beam roof. Spere trusses, more ornately carved and cusped than the rest, define what served as the dais or high-table area below, while detailed decorative carving is everywhere. A whole world of carved figures, beasts, birds and plants inhabit this roof, many of them evoking the atmosphere and spirit of the wild forest.

In general, medieval Scottish hall and tower roofs would have been of lesser span, complexity and decoration than the grand examples of Darnaway, Stirling and Edinburgh. That they might have nonetheless been impressive is borne out by the roof over the hall of the townhouse of the Knights Hospitallers in Linlithgow, West Lothian, which was demolished in 1883. Probably dating from about 1500, the detailing of the hall bears similarities to the nearby royal palace, while the arch-braced, double-collared oak roof may have also have been a miniaturised replica, its modest span of less than 5.5m conveying an undeniably grand effect. The timber may well have come from nearby Torphichen, the main centre for the Hospitallers in Scotland, and from where Edward I is known to have used oaks for the building of his peel at Linlithgow a couple of centuries earlier.

Unlike Norway or some English regions, Scotland has no known timber church-building tradition, and even surviving medieval church roofs are quite few in number and mostly fragmentary. Probably the most complete example is that over the nave of the Church of the Holy Rude, Stirling, which may date from about 1470. It is of tie-beam and king-post construction, each principal rafter being of double or flitched form trapping the through purlins, and below each tie-beam there is an extended wall-post and arched brace. St John's, Perth, retains parts of a tie-beam roof of comparable form, but just how typical these principal rafter roofs were is hard to determine. Probably the majority of the roofs over Scottish churches, towers and houses were of much

simpler common rafter form, perhaps typified, for example, by the double collar-rafter roof of c.1500 at Tullibardine Church, Perthshire.

Indeed, among churches and houses alike, variations on a basic collar-rafter theme appear to have characterised Scottish roof construction for much of the later medieval and early modern periods. An almost standard feature was triangulated ashlaring at the wallhead, sometimes beam-filled, and Scottish practice came to favour the setting of the vertical ashlar posts in front of the horizontal sole-piece, sometimes continuing downwards as wall-posts. Another variation on the collar-rafter theme involved the use of stilted or curved arch braces beneath the collars, mostly intended to serve as the underside or formwork for a boarded ceiling of mock barrel-vaulted form, but sometimes also for structural purposes, especially where, as in Angus, the outer roof covering was made up of heavy stone slabs. In such areas, the rafters tended to be more closely spaced than usual and were superimposed by open battens to which the slates were pegged, all of which remains visible in the roof of sixteenth-century Claypotts on the outskirts of Dundee. Elsewhere, the groundwork for the roof covering was mainly formed by close-butted or close-jointed sarking boards to which the slates were nailed or pegged and which, in the absence of purlins, provided some lateral rigidity. In houses comprising assemblages of ranges, towers and turrets, the conjunction of adjacent roofs and the creation of swept valleys and saddles on systems of interconnecting valleys, hip- and jack-rafters is usually a demonstrably later modification of eighteenth or nineteenth-century date, though recent examination of Newark Castle, Port Glasgow, Renfrewshire, has pointed to the possible use of valley-rafters between two juxtaposed ranges in about 1598. The bulk of the available evidence otherwise points to an original separation of adjacent roofs and the extensive use of

intermediate or valley gutters, no doubt lined with much sheet lead.

Illustrated in Italian sources, such as Palladio in the third quarter of the sixteenth century, and introduced into Britain by Inigo Jones (1573–1652), the trussed roof was a new form in which, characteristically, king or queen posts literally 'trussed up' the tie-beam over much increased spans. Just when the trussed roof was introduced into Scotland remains an open and unanswered question, but no seventeenth-century evidence of such a roof type has yet been uncovered. Roofs of hipped and shallow-pitched mansions of the Restoration and later eras that have so far been examined appear to have followed traditional structural forms, and present indications are that it might not have made its appearance here until the second quarter of the eighteenth century, perhaps initially and tentatively at the hands of the celebrated craftsman-turned-architect, William Adam (1689–1748).

Timber-framed houses are absent from the Scottish countryside, a lack that is explained in part by a social and tenurial structure that, unlike, say, England and Wales, did not embrace sufficiently numerous or substantial tenants and yeoman freeholders. However, in Scottish towns, as elsewhere in Europe, timber construction remained the norm for many centuries, despite increasing reliance on timber imports. Possibly emulating foreign fashion, late sixteenth-century Edinburgh witnessed a noticeable resurgence in the building of houses almost completely of timber on a substantial and expensive scale. Until their demolition in the last quarter of the nineteenth century, a few timber-framed and jettied buildings of this kind, evidently with masonry walling only at ground level and in the party walls, stood in and around the Lawnmarket. Drawings of one of these buildings showed that, in a small portion above a pend, it was wholly timber-framed and jettied at the back as well as the front, the only recorded occurrence of this feature in Scotland.

For the most part, as many contemporary visitors noted, many of the 'timber' houses were actually stone-built structures, with only galleries of timber, a type still exemplified by the restored timber frontage of the so-called John Knox's House in the High Street, Edinburgh. Dating originally from about 1570, its frontage is jettied in two stages and has small foregalleries set wholly in front of a load-bearing stone wall. Until its demolition in 1966, a more complete early seventeenth-century specimen of such a timber-fronted house with unrestored timberwork stood in the Watergate, Perth. A former town house of the first Earl of Kinnoull, it was a three-storeyed building with a circular stair-tower projecting into the street at one angle, the masonry walls of the upper two floors being set behind a galleried timber frontage which ran the length of the street elevation and partly encased the stair-tower. The framing, of oak throughout, was of independent construction at each floor level and preserved most of the original studs and mid-rails which served as sub-frames for window openings. Grooves in the sill beams pointed to an earlier form of panel infill, probably of timber boards or wattle.

Many early floors were of double- or triple-tiered construction, analogous in strength and flexibility to those designed for later industrial buildings with moving loads. This form of flooring usually incorporated a pair of runner beams carried on corbels set into the side-walls; these runners supported the main joists which were in turn notched to receive short transverse filler-joists. Wider spans would probably have had central rows of posts, which would be set on padstones and support longitudinal or spine beams.

A structurally important but hidden use of timber was as a foundation material. In Scotland as elsewhere, vertically set piles and horizontal grids or frameworks

10

of timber known as 'branders' (or as 'starlings' when they were rubble-filled), were commonly used in river bridge foundations, particularly in tidal rivers. Similar substructures were also created for buildings founded on light and sandy soils (as in the case of the lofty late medieval Bishop's Palace in the centre of Dornoch) or on boggy ground, a technique that is best represented in Scotland by King's College, Aberdeen, founded in 1505. A mid-seventeenth century topographical account records that, as the college was constructed on marshy ground, the expensive and time-consuming process of laying oak-beam foundations had to be undertaken.

In the course of the nineteenth century, trussed timber roofs came very much into their own, especially among large structures with clear spans of upwards of 8–10m; that is, industrial workshops and warehouses, as well as grand mansions and

The sanatorium at Glen o' Dee is a typical if sadly dilapidated timber-clad building, built following improvements in rail transport in the nineteenth century

The last timber mainline
railway viaduct in use
in Britain, standing
impervious to the winter
storm: Aultnaslanach, Moy,
Inverness-shire

13

churches. Associated developments in sawmilling
in town and country fed these burgeoning demands
for machine-sawn timber which, conspicuously in
north-eastern Scotland, also came to be applied
extensively as external cladding across a vast range
of building types in the region, the wholly timber-
framed Swiss Cottage of 1835 at Fochabers being
among the earliest and most exotic. Timber-clad or
weather-boarded structures were common on the
railways, especially in those same parts of northern
and north-eastern Scotland served prior to 1923 by
the former Highland and Great North of Scotland
Railways. Paralleling developments on the American
railroads, the Scottish railways also became heir to an
ancient timber or timber-and-stone bridge-building
tradition. This was well exemplified by a wooden
cantilever bridge on stone piers which used to carry
the railway over the River Forth at Stirling, while
the line between Aviemore and Inverness that was
opened in 1897 retains a trestle viaduct composed
entirely of pitch-pine baulks with iron-clad joints.
This bridge at Aultnaslanach, the last of its type on a
main line in Britain, owes its survival to the fact that
its foundations were long considered to be better
suited to the surrounding boggy ground than those
of a heavier bridge of metal or masonry.

In the recent, ecologically conscious decades at the turn of the twentieth to the twenty-first century, timber has reacquired something of its former strength and character as a building and cladding material, informing, inspiring and shaping an entirely new generation of designs in which its use has been underpinned by a greater appreciation of its behavioural properties and by the application of science and technology to its effective all-weather performance. Across almost the entire country, from Sleat on Skye to North Berwick in East Lothian via, for example, Glencoe, Aberfeldy, Dundee and Edinburgh, modern Scotland has already created a distinguished corpus of timber designs that is worth both celebrating and developing further.

Timber is once again becoming a widely used material in modern Scottish architecture: Maggie's Centre, Dundee

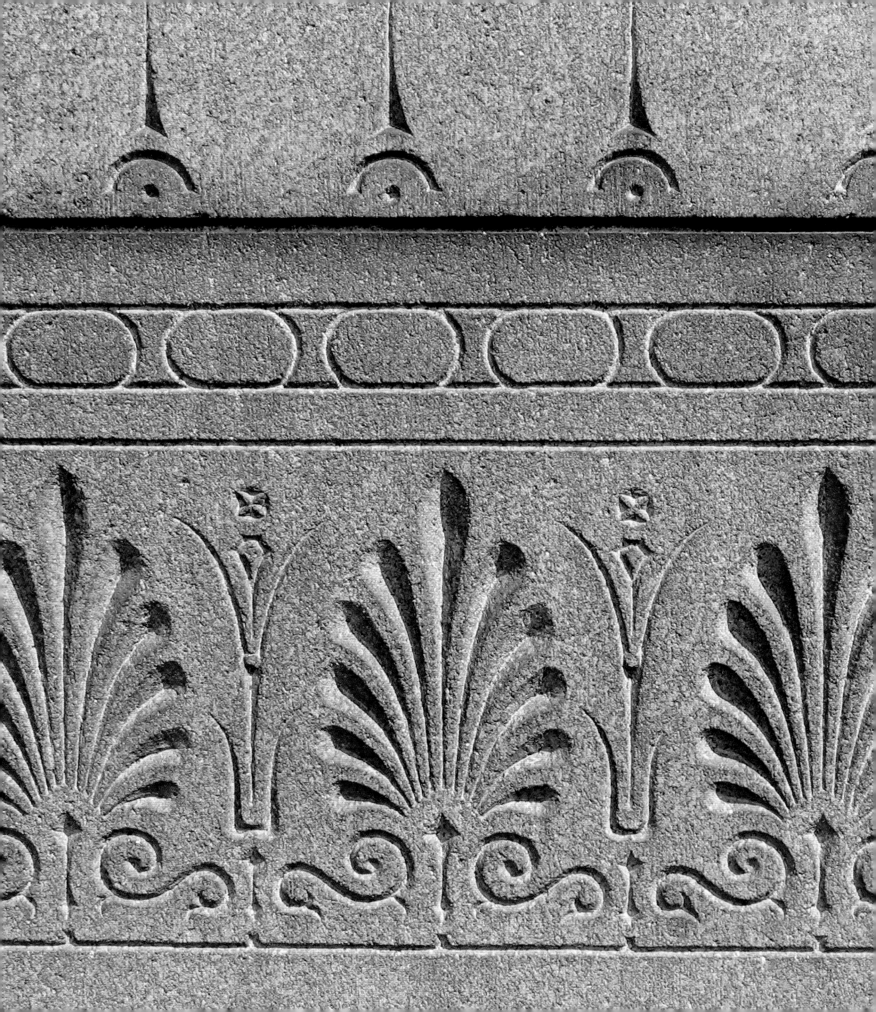

CHAPTER 2

STONE

Ingval Maxwell

...the link between architecture and geology is inescapable.

Introduction

Every structure that has ever been built has been constructed from materials that have either grown on the surface of the earth or have been excavated from beneath it. In that process the link between architecture and geology is inescapable. It follows that, over the last 5,000 years, Scotland's traditional masonry buildings have naturally and inevitably displayed the inherent characteristics of the sedimentary, igneous and metamorphic rocks from which they were built.

Rock and building stone

Although Scotland is a country created from a complex variety of rock types, sedimentary rock is the predominant building material in the Northern Isles, around the Moray Firth, and across the central belt and southern Scotland.

Sedimentary stones, such as sandstone, were produced by natural weathering, disintegration and decomposition processes which affected pre-existing rock. The eroded particles, transported by wind or water, were deposited elsewhere, usually in stratified layers. Over time these loosely formed layers were buried, compacted and bonded by being subjected to high temperatures and pressures deep within the earth's surface crust. Geologically bound, these processes produced high-quality building blocks as the consolidated strata re-emerged at the surface.

Mainly consisting of calcium and magnesium carbonate, limestone primarily originates from the geological conversion of

17

ABOVE: Stone quarrying was at one time carried out on a truly massive scale in Scotland
© NERC BGS

OPPOSITE: Awe-inspiring stone circles were some of the earliest uses to which the material was put in Scotland

accumulations of marine deposits. Limited in its use as a building stone in Scotland, limestone is the key material used in the production of traditional lime mortars, renders and plasters. Found in pockets throughout the country, concentrations of limekiln and quarry workings are located in Charlestown, Fife, Lismore in Argyll, the Lothians and Dumfriesshire.

Whilst these sedimentary rocks were originally produced by processes that operate mainly at the earth's surface, igneous rocks, such as granite, are formed by the solidification of hot, molten or partially molten rock material (magma) which has been extruded from below the surface crust. Characteristically, their solidified appearance is crystalline and uniform, and this is coupled with considerable strength, affording a robust, stable and resilient building material.

As a result of variations in temperature, pressure and mechanical stress, metamorphic rocks result from the geological alteration of pre-existing rocks in consequence of changing environmental conditions and minerals. The pre-existing rocks may be of igneous, sedimentary or metamorphic origin. The alterations are generally brought about by either the intrusion of hot magma into cooler surrounding rocks or by large-scale tectonic movement that creates pressure and temperature changes. In the process, the original minerals responded by reacting with one another to produce new stable minerals, resulting eventually in durable, hard-wearing building stone.

Significant extrusions have determined that granite structures are built in the north-east and south-west of the country, whilst metamorphic slates, used for roofing purposes, have primarily been obtained from quarry sources around Ballacullish and Easdale in Argyll, the Macduff area in the north-east and along the Highland Boundary Fault. Sources of greywacke (whin) in the Southern Uplands, together with schist and gneiss in the north and west, have also been used for building.

The most recent geological episode, the Quaternary period, is best described as a time of numerous climatic changes which fluctuated between dry and humid conditions. These cycles had a dramatic effect on the landscape, resulting in a wide variety of loose deposits being laid down on the surface, which could be readily quarried. Small rock or soil particles are generally referred to as sand, of which quartz is by far the most common element. Quartz is comparatively hard, cannot be readily worn down, is generally insoluble in water, and does not decompose. Through the preparation of cement and concrete, the modern construction industry uses sand in a multitude of ways, but its use can be traced back to the Roman occupation of Scotland, when lime mortar, rendering and plastering techniques were introduced.

Combined, these basic materials quite literally provided the building blocks of Scottish architecture. Over the period of human habitation in Scotland, some 12,000 building stone quarry sources of different sizes and outputs are estimated to have been opened across the face of the country, with over 750 locations providing good-quality materials to satisfy the peak of masonry building that occurred during the mid-to-late nineteenth century.

Using the materials

In prehistoric times, awe-inspiring megaliths were meticulously placed on the landscape in the form of stone circles, stone alignments and settings. Speculatively we relate these sites to an understanding of astronomy, seasonal changes and meteorology, but their true purposes are still unknown. The many sites, including the great stone settings of Brodgar and Stenness in Orkney, and the alignment of the principal setting, together with its satellite sites, at Callanish in Lewis, are impressively commanding in their visual impact. The certainty with which the megaliths were placed in their foundation, and their orientation, can only hint at the understanding that early man had of how to use stone in this significant manner.

Developing the same prehistoric skills which created domestic settlements like Knap of Howar and Skara Brae in Orkney, Scotland's builders reached their zenith in drystone construction 2,000 years ago. Capable of adapting their approach to the use of a wide variety of rock types, these constructions, of concentric walls built in the form of a conical tower, present a quality of build that has never been surpassed. There was complete mastery displayed in how the stone was chosen, placed and set. Broch construction shows free-flowing plan form, gentle corbelling of curvilinear shapes, and the use of large slabs to create partitions, lintels and furniture. This serves to emphasise how competent Scotland's early masons were in utilising available geology.

Drystone build, sometimes combined with the use of clay-based infill packing, also illustrates an awareness that the structural strength of what was built lay in the need for each successive course of masonry to directly bear on the upper face of the underlying course. In doing this, builders created stable walls, transmitting the weight of the stonework directly to the foundations. This technique allowed

ABOVE: Building in un-mortared stone reached its technical peak with the construction of the massive brochs throughout Scotland

OPPOSITE: Artistry and engineering combined to create beautiful monuments such as the Kildalton cross, Islay

the building to flex as climatic conditions dictated, and this perhaps helps to explain why so many structures from this era survive today.

Art and engineering

Embryonic artistic skill in working stone was initiated with the prehistoric technique of incising 'cup and ring' and linear marks on the surface of exposed bedrock and some large loose boulders. The second significant period of artwork in stone had to wait until the sixth century AD with the creation of the early Pictish symbol stones. Here, with a lineage and skill that had become increasingly sophisticated by the ninth century, the highly ornate and exquisitely carved monoliths such as Sueno's Stone, Forres, Kildalton Cross, Islay, and St John's Cross, Iona,

bring another peak in the manipulation and use of sedimentary, igneous and metamorphic stone in Scotland. Offering an interlinked mixture of pagan and Christian symbolism, the capacity to conceive, quarry, transport, carve and erect such statements of truly epic proportions was quite remarkable.

In the Middle Ages, this artistic process continued to develop whilst at the same time becoming inextricably integrated with an emerging ability to exploit structurally the full potential of the properties of stone. Influenced by the spread of the various religious orders, from the twelfth century a variety of architectural forms, carved details and styles were promoted in the construction of great abbeys, cathedrals and monastic ranges. Stone, slate, sand and lime were the essential ingredients,

and their successful integration provided the key by which engineering, artistic abilities and constructional challenges were solved. Scotland's link with mainland Europe was well and truly established, and the quality, style and extent of its architecture influenced accordingly. Assured by its ready availability, accessibility, and relative ease of working, sandstone became the building material of choice.

Robustness and strength

As power shifted from the ecclesiastical to the secular, internal civil disruption and unrest overtook stability. Consequently, from the thirteenth century onwards, a necessary surge in rapid defensive building occurred with the erection of innumerable medieval castles. Reflecting the immediacy of construction, a broader range of building stone types, including granite and whin, from a variety of new quarry sites, was used.

The need was to create strong, thick, defensive walls, and these required the acquisition of large volumes of stone and mortar. But the requirement for building economically, then as now, was also important. Why transport weighty stone unnecessary distances if it could be found on the doorstep? Why manufacture large quantities of costly lime mortar mixes if greater volumes of stronger stone could instead be used in creating the walls? Why throw away any of the stone that was hard-won from the quarry?

Sensibly, quarried sources of stone, clay and lime for mortar were inevitably procured from the immediate vicinity. Elcho Castle near Perth and Craigmillar Castle on the outskirts of Edinburgh are two examples of where this functional synergy can be readily seen. In both cases, the defensive rubble stone walls sit directly on bedrock, with evidence of associated quarry working immediately below.

A closer examination of the way in which rubble

23

ABOVE: Blocks of varying dimensions form the mass from which massive stone fortresses such as Caerlaverock derived their strength

OPPOSITE: Some of the finest medieval Scottish stonework is to be found on ecclesiastical buildings, such as this fine ribbed vault

walls were built confirms that everything extracted from the quarry was used. Single blocks of large dimensions were reserved for structural lintels, sills and dressings around openings; while a multitude of differently sized stones, including numerous smaller pinning stones set around the large rough blocks, made up the general wall face, thickness and volume. The more stone that could be incorporated into the wall, the stronger and more durable it would be, whilst at the same time reducing the amount of mortar required to keep it bonded together.

However, the natural properties of the different stone types, their bed thickness, strength, cross-fracturing and available block sizes, also dictated the available dimensions of stone to span openings, create sills and corbels, and provide steps. Where this happened, different architectural solutions were adopted to achieve similar ends. Arches spanned openings instead of flat slab lintels; steps were created from a variety of stones rather than single pieces; and smaller-sized blocks were tooled with common mouldings to make up continuous runs. Geological properties predetermined much of the eventual appearance of what was built.

Stability and refinement

The country became more stable during the late medieval period, and new fashions and influences manifested themselves in architecture, as is perhaps best revealed in the Upper Courtyard of Stirling Castle. Here, on four sides of the square, a different architectural style is evident on each of the buildings that was constructed during the course of the late fifteenth century and the sixteenth century. Scotland's medieval European links are more than adequately displayed, and the scene was set for the nobility to emulate these models in the construction of their own buildings.

The desire to be architecturally fashionable grew.

From the eighteenth century, other international models were widely promoted following the publication of a range of books and influences emanating from the various Grand Tours that the nobility, gentry and sponsored professionals embarked upon. Bringing back a unique insight into the design and quality of ancient and classical architecture, the refined details of marble, limestone, brick and stucco were readily adopted, using indigenous Scottish stone. In the construction industry of the time, as prosperity grew, so did innovation. A significant growth in quarrying, transporting, hewing, machining and carving stone

followed. The diversity of Scotland's geology was truly reflected in the expansion of its urban and rural architecture.

Craig's 1736 plans for Edinburgh's New Town set a trend that many were to follow. The grand, classically influenced and integrated palatial designs of the street elevations set new standards in planning, design and architectural detailing. The result was a unique combination of quality architecture and cost-effective building. Ashlar principal facades were created in parallel with the more economic use of rubble-stone side and rear walls, and internal partitions.

This pattern was to be repeated elsewhere, and as demand continued to rise, by the mid-nineteenth century many local supply quarries had either been worked out or built on. Sources of stone had to come from further afield. The needs of Glasgow and Edinburgh were initially aided by the construction of the Union and Forth and Clyde canals. Subsequently, the growth of a more comprehensive railway network enabled the transportation of stone to a much larger market. Imparting their own distinctiveness in the appearance and quality of the available material, large productive sandstone quarries became profitable. With much improved transportation, stone supplies from quarries at Ravelstone and

Taking advantage of nature: building around bedrock, Dirleton Castle

LEFT: Large yet skilfully worked masonry employed to construct Edinburgh's New Town

OPPOSITE: Renaissance influences captured in stone: Crichton Castle

Hailes near Edinburgh; Humbie and Binnie in West Lothian; Plean and Polmaise near Bannockburn; the Grange and Cullaloe in Fife; Ballochmyle in Ayrshire; and Gatelawbridge, Corsehill and Locharbriggs in Dumfriesshire, amongst others, satisfied the growing need. However, by the end of the nineteenth century, as demand continued to grow, both Glasgow and Edinburgh were ultimately drawing in stone from as far afield as the English Midlands and the north of Scotland. Previously accessible local quarry locations subsequently became lost and used for building, occasionally hosting such landmark features as the Scott Monument and Glasgow's Queen Street Station.

International opportunities also emerged. During the late nineteenth century, as trade links with North America developed, Scottish building stone was frequently exported from the Solway and Clyde ports as ballast on returning ships. As a result, buildings with a distinctly Scottish flavour can still be seen along the eastern seaboard of America and Canada.

Significant developments also occurred in quarrying and exporting granite. From 1764, Aberdeen granite was being transported by sea and used to pave the streets of London; by 1795, it was being used in the construction of Portsmouth's docks. Similar opportunities were presented to the granite quarry masters in Galloway. In addition to creating the granite towns of Dalbeattie and Castle Douglas and nearby villages, the interest of the Liverpool Dock Trustees and others in the quality of the stone ensured much activity in the seaborne supply of blocks for the building of a variety of docks including London's Embankment during the first half of the nineteenth century.

Quality paving stone and durable stone roofing slates were also produced and widely exported from the readily worked Caithness sources and the Angus

OPPOSITE: Drystone walling has enclosed many thousands of miles of Scotland's fields since the period of agricultural improvement: Orkney

quarries at Carmylie. The former is still in production, whilst the latter, having accommodated the county's needs since medieval times, is currently dormant. Caithness flagstone paving has found use throughout Scotland and as far afield as mainland Europe, India and South America.

Across the broad geological base, Scottish stone-working was taken to new heights with the introduction of numerous Victorian innovative industrial developments in pumping, machine sawing, lathe turning and the mechanical planning, cutting and dressing of stone. As a result, productivity greatly increased, as did the range of outputs and markets that were served.

Changing the landscape

With a rising urban population, increasing demand for stone was also placed on the rural economy. The pace of land enclosure during the agricultural improvements of the late eighteenth century and early nineteenth century required the fields to be cleared of loose boulders and stones so that they could be worked more productively. In the first stage of improvements, some of that material was consumed by making drainage ditches and boundary walls, but much was also used to create rudimentary, but better organised, farm buildings, farmhouses and cottages. Much advice and practical guidance followed, and published material by various improvers illustrated how to enhance these initial developments. This led to a significant rebuilding programme to create more functionally effective agricultural buildings during the middle part of the nineteenth century. Numerous steadings, farmhouses and cottages from that period still survive today, albeit under increasing pressure to be modernised once more.

Pressures and change

More modern construction materials started to make an appearance during the nineteenth century. Following its invention in the 1840s, the 'benefits' of Portland cement-based mortars, and concrete in particular, made considerable inroads in construction, with a watershed date of c.1920 as the use of traditional building techniques gave way to modern methods. Cement could be used in all weathers, was more controllable in its application and, once suitably reinforced, was more effective in its load-bearing capabilities. Being hard, strong and durable, in the early part of the twentieth century cement-based renders were regularly applied over traditionally built stone walls in the misguided belief that they kept the rain out. Likewise, general universal adoption of cement-based mortars almost relegated the use of lime to history. But the inherent strength and hardness of cement-based mortars and renders were also to create difficulties, although the full extent and significance of this was not realised until much later. With hindsight, the wholesale adoption of cement-based solutions for traditional buildings has been shown to be misguided. The use of cement has triggered damaging decay mechanisms that currently affect numerous stone buildings.

Stone also became less used as a structural load-bearing material in its own right, and by the 1960s, designers had resorted to specifying its application in building works as an aesthetic device, pinned or glued to walls in thin sheets. Change was pending, and by the 1990s there was an awareness of the benefits of using lime technologies in the repair of traditional buildings. This reawakening has provided the base upon which a greater understanding of the quality and benefits of traditional stonework has been reached. Equally, it has been recognised that much can be learned from studying, analysing and adopting the lessons of what was previously, and intuitively, understood in the construction of Scotland's masonry fabric in the past.

CHAPTER 3

BRICK

Moses Jenkins

. . . brick is an ancient and integral part of our building tradition.

If we think of traditional building materials in Scotland, brick is perhaps not one which would immediately spring to mind. However, the use of clay bricks as a building material has a long tradition in Scotland, and they have formed an integral part of our built heritage for a considerable time. Bricks have been used as a building material around the world for at least 5,000 years. The first recorded use of them in Scotland is in the first century AD, following the Roman invasion, for example at Newstead, Roxburghshire. During the period of industrialisation there was a great expansion in both the use and manufacture of bricks in Scotland, and they are utilised in a significant part of our built heritage. Brick continues to be an integral part of Scotland's building tradition to this day and, with a history dating back 2,000 years, it is a material well worth including in any work about traditional building materials.

Brick can be found as a building material throughout the length and breadth of the country. Brickworks were established wherever there was a suitable deposit of clay and a demand for bricks. Evidence of their existence can be found from Terally in Wigtonshire, Brora in Sutherland, Errol in Perthshire, Craigellachie in Banffshire, Kilchattan in Bute and Garabost in Lewis. Brick was used as detailing around windows in Galloway, to build warehouses in Aberdeen, mills in Dundee, miners' rows in Ayrshire and cottages in Sutherland. It was the material used to build lighthouses in Islay, bridges in Deeside, distilleries in Fort William, tenements in Glasgow and factories in Paisley. Despite this wide geographical spread, brick usage in Scotland was concentrated in the industrialised central belt. It was in this area that the greatest deposits of the raw materials used to manufacture bricks were found, and it was where the greatest demand for building materials existed.

The basic raw material of brick is clay. There are two types of clay-based material which have been used throughout the centuries in the making of bricks in Scotland: pure clays, which were extracted and used to form bricks; and red blaes, or colliery shale, which was clay mixed with other materials extracted during mining operations. These materials make bricks of differing quality and for differing uses. Most clay deposits in Scotland originate from carboniferous mudstones deposited in the wake of the last Ice Age, although other geological deposits have been used in brick making; for example, a seam of Jurassic-age clay was utilised

by the Brora brickworks. Carboniferous mudstones vary in quality, with some being too high in carbon and sulphur for use in brick making. Such carboniferous clay deposits were often found interspersed with coal fields of the same geological age, meaning clay and coal extraction often went hand in hand.

Shale and blaes were not suitable for manufacturing high-quality facing or engineering brick but were instead used to produce common or 'colliery' bricks. Such common bricks were often used for internal partitions or were harled and generally kept out of view. This was for two reasons. First they did not have the consistency to withstand the weather because of inconsistencies

Brick can be found as a construction material throughout the country: here it is used to build a row of cottages in Sutherland

Brick is particularly common in areas of coal production, the two industries often going hand in hand due to an overlap in natural resources: Lady Victoria Colliery

in the raw material. Second they possessed a variegated pattern rather than the deep red of a purer clay brick.

Clay which was extracted and not mixed with shale material came from a number of sources. It was made into facing or engineering bricks and formed a strong versatile brick. Such bricks were often very strong, as the high temperatures involved in the firing caused vitrification of the face. Suitable deposits to create these high-quality facing bricks were found in Lanarkshire and Ayrshire in particular, although most brickworks outwith the central belt had access to suitable clay to produce such bricks. Other types of facing bricks were manufactured, including white clay (often found as detailing around windows and other parts of buildings) and

glazed bricks, which were used either as decoration or in situations where a high level of cleanliness or light exposure was required.

The highest-grade clay was manufactured into engineering bricks. These were denser than common or facing bricks and had to meet certain standards of compressive strength and water resistance. They were used to build structures such as bridges, tunnels and lighthouses, or in elements of a building that required greater strength than common bricks could provide. High-quality fire clay was used to manufacture refractory bricks for use in furnaces and other industrial applications but, although Scotland was a world leader in the industry, a discussion of such bricks is outwith the scope of this chapter.

An eighteenth-century brick-built farmhouse constructed using locally produced handmade brick

In the early period of brick manufacture in Scotland, clay was simply shaped by hand in a mould before being allowed to dry and then fired. This firing sometimes took place in a clamp, a temporary construction of unfired bricks which was dismantled after firing and could be erected near the clay source. A temporary kiln often constructed of turf was also sometimes established on the site of building works. We know from contemporary accounts that these kilns could hold around 25,000 bricks and took 8 or 9 days to complete burning. The normal process in brickmaking was that clay would have been dug throughout the winter months, and then a skilled brick maker, who travelled with his staff from site to site, would come to mould and fire the bricks in the summer.

Although itinerant workers continued to be used into the nineteenth century, the eighteenth century saw an increase in the establishment of permanent brickworks. William Adam started a brickworks in Kirkcaldy in 1714, but other works had been begun before then – for example, at Leith in 1709. Throughout

the eighteenth century, new works continued to be established, such as Cupar brickworks in 1764, and Prestongrange and Portobello, which were exporting bricks as far afield as the West Indies in 1779.

With the advent of these large-scale operations came developments in both moulding and firing technology that made the manufacture of bricks more efficient. Pug mills (a mill used to process clay before moulding) were developed, which gave the clay an even consistency and helped remove air. This new technology allowed more advanced techniques of brick manufacture, such as the extruded clay process. Extrusion involves the formation of a column of clay pushed through a die. The resulting column of clay is then cut, using wires, before the bricks are dried and then fired. This process was only suitable for clays with sufficient plasticity to be forced through the die. The invention of the brick press (a machine for forcing clay into moulds mechanically rather than by hand) led to the development of the stiff plastic process. Where clay was not suitable for the extruded method of preparation, the stiff plastic method was

employed. In this process the clay, shale or blaes is ground, crushed and mixed with other substances using a grinding mill or pug mill to form a stiff paste, which is then pressed into a mould before proceeding to firing. This was the process by which colliery or composition bricks were formed.

The advent of the movement for agricultural improvement saw the development of a great number of small brickworks for use by tenants on estates. These brickworks would often see tenants digging and moulding their own clay before being allowed to use a communal kiln. Examples of these brickworks and methods of working are evident at Blairdrummond estate and at the Stix brick and tile works in Perthshire, demonstrating that small-scale local production still had an important part to play in Scotland's building tradition despite the rise of the larger works of the central belt.

Developments in kiln technology mirrored those in the moulding process. Kilns for firing bricks can be split into two broad types: intermittent and continuous. The simplest form of intermittent kiln is the up-draught kiln, which consisted of one chamber and was fuelled by coal distributed throughout the bricks in the kiln or, more commonly, in fireholes at the base. They suffered from poor fuel efficiency and heat loss, and sometimes resulted in varying degrees of firing. Up-draught kilns were superseded by down-draught kilns, which permitted more uniformity of heating and produced a superior product. They were sometimes fuelled by oil or gas, but coal was most common. In 1858 the first continuous kiln came to Scotland. This was the Hoffman Kiln, and it proved to be another significant advance in the technology of brick manufacture. A continuous kiln worked on the basis that, as one chamber was being fired, unfired

Advances in technology meant that bricks in a variety of shapes and colours could be produced, allowing ever more complex decoration

LEFT: Stanley Mill, Perthshire: one of the earliest industrial buildings to use brick in its construction in Scotland

OPPOSITE: Due to its heat-retention properties, an early use of brick in Scotland was the formation of garden walls

'green' bricks stacked in the adjacent chamber were dried by the exhaust gases from the one being fired. The firing area then moved to the next chamber, and so on. After an adequate period of cooling, the bricks from the chamber which had previously been fired were removed. This process was ideal for the production of composition colliery bricks made from colliery shale or blaes. Later developments in kiln technology led to further advances in efficiency in production. These included the Belgian Kiln, the Staffordshire Kiln and the Dunnachie Kiln, which was developed by a Scot, James Dunnachie.

Throughout the 2,000 or so years in which brick has been used in Scotland, it has seen many different applications. Brick has been used in Scotland from as early as the first century AD. The Romans brought brick-making technology to this country and utilised local clay to make bricks at a variety of sites throughout Scotland. Locally manufactured bricks have been found in Newstead Fort, Balmuildy Fort, Old Kilpatrick Fort, Castledykes Fort and Bothwellhaugh Fort. The Romans primarily used their thin, flat bricks for supporting hypocaust (underfloor heating) systems in high-status buildings and bath houses. Brick was then used only sporadically up to the seventeenth century. As brick began to assume

a greater prominence as a building material in the seventeenth century, it was often used to build garden walls because of its heat-retaining qualities. Master gardener John Reid, in his work *The Scots Gard'ner* published in 1683, recommended brick for use in garden walls: 'as for walls, brick is best'. In 1643, it was recorded in the minutes of the Privy Council that Tobaccos Knowes received a patent 'for the making of bricks under several conditions'. Examples of early brick garden walls can be seen at Megginch in Perthshire, at Drummond Castle and at the Pineapple in Dunmore. Brick was also used to build chimneys and other structures: for example, chimneys at Sanquhar Castle in 1688, floors in Edinburgh's Smeatons Close, and at the Cromwellian Citadel in Inverness a breastwork which was noted by a local minister as 'lined within with a brick wall'.

It was during the eighteenth century that brick began to be used on a more significant scale in Scotland than in previous centuries. Brick retaining walls and internal partitions were constructed at the forts that were built following the Jacobite rebellion. Brick also took on a prominence in far greater undertakings. With the demand for bricks on the increase, they came to be used in high-status buildings such as Inverary Castle and Gordon Castle, and by prominent craftsmen such as William Adam. The Duke of Argyll employed several brick makers under the direction of Adam to make bricks for his remodelling of Inverary Castle in the mid-eighteenth century. Other small, free-standing structures were built of brick during this century, including dovecotes, washhouses and pavilions. There are examples of some of these at Buchanan, Stobhall and Dunrobin castles. As mentioned earlier, brick also came to be used increasingly when the movement for agricultural improvement at the end of the eighteenth century began to create better living conditions. One observer described how on an estate 'As soon as they have cleared a little ground they build houses of brick. There are now [1791] 69 brick houses substantially built using lime.'

The nineteenth century saw brick used to construct large free-standing structures in a way that had previously been rare. Technological advances allowed the more efficient and economic production of bricks throughout the nineteenth century, and the use of colliery shale as a raw material led to an explosion of manufacture.

Brick and cast iron often combined to form striking industrial buildings, such as the sugar warehouse, Greenock

OPPOSITE: Probably the finest example of brickwork in Scotland, the polychromatic splendour of Templeton's Carpet Factory

In 1802, 714 million bricks were manufactured in Scotland, but by 1840 this had jumped to 1,725 million. This growth in manufacture was matched by a growth in demand as bricks were used to build the structures that powered the Industrial Revolution. These grandiose temples of industry used massive quantities of brick in their construction, with a number of fine examples surviving to this day. Houldsworths Mill, built in 1805 at Anderston, Glasgow, was one of the earliest large buildings to be built entirely of brick in Scotland. Brick-built mills and factories exist in Cambusbarron, Stirlingshire; Barrowfield Weaving Factory, Glasgow; and the Anchor Mills in Paisley. A particularly fine example is the sugar warehouses in Greenock built between 1882 and 1886 to designs by Walter Kinipple and constructed in four red-brick sections, with arches and pilasters in yellow brick. Broadford Works in Aberdeen, constructed in 1912, was certainly one of the largest brick-built structures in Scotland. It was known locally as the Bastille because of its imposing size. The most striking example of the use of brick in an industrial building is surely Templeton's Carpet Factory in Glasgow. This was built to a design by William Leiper, and was supposedly modelled on the Doge's Palace in Venice. It uses an array of polychromatic glazed bricks made by the Cleghorn Company of Glasgow to create a truly beautiful building.

Other brick-built industrial buildings include a water tower in Dalkeith, designed by James Leslie and built in 1879, which featured cream brick decoration; and the Alloa Glass Works cone, constructed from local bricks in 1824. It was not only the large-scale industry of Scotland's central belt that made use of brick in the construction of its buildings – rural industry made extensive use of the material as well. Kilns were made of it at farms such as the one at Preston Mill; at Milnholm Fish Hatchery, which has unusual yellow brick walls; and at Castle Douglas

Cattle Mart, a polychrome cream brick and red sandstone octagonal hall.

Glenlochy and Dallas Dhu distilleries (both built in the 1890s) were constructed using brick, and almost every distillery used the material for their chimneys. Nineteenth-century railway stations, such as at Fort William (built 1894), and Aberdeen station, which has a brick vaulted substructure supported by columns, were also frequently brick-built .

Churches, often viewed as a bastion of stone traditionalism, were increasingly built of brick from the mid-nineteenth century onwards, particularly those outwith the established Church. Examples of these include Morebattle Free Church, Roxburghshire, built in 1866, which incorporates decorative brick detailing, and St Matthew's Roman Catholic Church in Rosewell, built by Archibald MacPherson in 1926. Many churches by Jack Coia are constructed in brick, such as those at St Anne's, Dennistoun, and St Peter's, Ardrossan. Perhaps the most striking brick-built church is St Sophia's in Galston. It is modelled on a church in Constantinople and was built in 1885. It is another striking example of decorative brickwork being used to great effect in Scotland.

Brick was also used for engineering. Bridges were often built with brick reinforcement, as can be seen at Dunglass Railway Bridge, Berwickshire, built in 1840; at Ruthrieston Bridge, Deeside, where brick vaulting was employed; and on the piers for the Tay Railway Bridge, erected in 1887. The viaduct at Inchbrayock, Montrose, opened in 1878, is constructed entirely from brick. Other engineering works utilising brick include tunnels such as the Killiecrankie railway tunnel. Lighthouses, too, were frequently brick-built: examples can be seen at North Unst and the Butt of Lewis.

Despite these many and varied applications, the most significant use of bricks in Scotland has been for housing. Following the Industrial Revolution,

OPPOSITE: An early example of polychromatic brickwork: Rosemount buildings, Edinburgh

BELOW: It was not just the machinery of industry that was housed in brick buildings, so too were workers throughout the country: Newtongrange

brick was used extensively in the construction of workers' homes. Whilst many examples have been lost because of subsequent redevelopment, brick-built rows of houses for miners and iron workers existed throughout the central belt – for example, at Annathill, Lanarkshire; Kilwinning, Ayrshire; and Fishcross, Clackmannanshire. One row of miners' houses was described in a report on the condition of miners accommodation as 'ten single apartment houses, five on each side built back to back. They are built of brick. The house measures 11 feet by 10 feet.' Although long terraces of brick houses are rare in Scotland, examples do still exist. Most of these were built where brick was available locally, such as at Brora and Prestongrange, where there are terraces built from bricks produced at the nearby works. An early example is the Rosemount Buildings in Edinburgh, built in 1859 to house 'the better class of mechanics'. Brick also formed the internal walls and other unseen structural elements of many buildings, and it was common to use harling to protect external brick walls from the weather. The last two mentioned techniques have ensured that the majority of bricks in the Scottish built heritage remain largely unseen: this, more than anything, has ensured that their place in our built heritage has been unappreciated. Nevertheless, as this chapter has shown, brick is an ancient and integral part of our building tradition.

CHAPTER 4

EARTH

Chris McGregor

Earth, perceived as a material used as a necessity out of poverty, should be looked upon as a wonderful, eminently sustainable material

Although stone is generally perceived to be the principal building material used in traditional Scottish construction, earth is now recognised as having predominated throughout most of Scotland's history.

There is a range of earth construction to be found over the whole country. The techniques are varied, but the consistent element is earth, which is the most sustainable building material used by man. Owing to the complexity of Scotland's geology, there is no overriding earth building technique that can claim to represent the whole country.

Naturally occurring earthen building materials

One of the simplest ways of using earth for building is to cut material that is naturally reinforced with vegetable fibre. Turf with thick fibrous roots keeps its shape during the building process, if cut and built at the right time of year. The traditional builders understood this well, and we still have examples of this form of construction. We see many ruinous buildings in the countryside today which appear to have lost their gable ends: many of these dwellings and farm buildings may have had turf gables, or caber and mote gables.

Turf gables were usually built off a flat wallhead. The turf, or in some cases peat, was often cut in parallelogram shape and laid in a herringbone pattern. It is suggested that this pattern might allow the blocks to bond more tightly, as the self-weight of the turf allows the wall to settle.

Lime pointing was only used in limited areas, either as a

statement of status or, as in the likes of a milking parlour, as a hygiene agent. Many properties had their own lime kiln. However, virtually all of the lime burnt was intended for the fields, to break up the local boulder clay and improve their productivity.

Entire buildings were constructed of turf or peat in a similar way. In most cases these walls would be built off a stone base course. Some had a drystone inner wall cladding, particularly when used for storing animals, to provide a wearing surface that would protect the dry earthen walls from the cattle. Later, as with the Blackhouse at Arnol, and most buildings of this type of construction, the walls were clad both internally and externally with a drystone or earth-mortared face. In the case of Lewis, it was a direction from the estate. It perhaps provided a more permanent appearance to the dwellings, but the external cladding of earthen walls can prove problematic. The floor is earthen; traditionally the

sheep would be herded into the house to trample the clay.

In this treeless environment, timber was a precious material; thus the span was kept to a minimum by bringing the roof timbers to the inside face of the wallhead. As a consequence of using this construction technique, it was essential to waterproof the wallheads. These were waterproofed by applying blue clay, usually found at the bottom of the peat bogs. A layer of turf was placed over the clay to provide it with some protection.

In chimneyless dwellings, such as the blackhouse, the turf was omitted from a section of the roof to one side of the hearth, in order to allow the smoke to percolate through the thatch. This small section of smoke-blackened straw had to be re-thatched more frequently than the rest of the roof. However, nothing was wasted. This thatch would go back onto the field to fertilise the crop of potatoes or wheat,

Blackhouse, Arnol

which would provide the fresh straw for the repair or replacement thatch for the dwelling.

Peat or turf is still used as a fuel today, cut from the local peat bog and stacked to dry out over the summer months. It is formed into a large peat stack at the dwelling before being put on the hearth.

Turf was traditionally used as an undercloak to most thatched roofs in Scotland. The rough hewn timbers of most vernacular buildings could be smoothed out by the application of turf. The turves would be laid onto the roof in the same way that slates would be used to allow the water to run off the roof. Turf could be laid face-up or face-down. Usually where it was a base coat, it would be laid face-down. Where turf is used as a wallhead material or a ridge covering, it is laid grass-side up so that the turf can remain alive. The thatch would be laid over or dressed (stobbed) into the turf.

The range of thatching materials was considerable.

Corse Croft gable

The most abundant locally available material became the thatch of choice. However, some of these thatches were more robust than others, and people might travel further afield to find a more durable thatch.

At Corse Croft, the entire house is constructed of locally sourced turf .

Interior, Blackhouse, Arnol

Mudwall Construction

Priorslyn Barn

This structure demonstrates the technique used in this area. The earth walls are built up in layers of 150mm (6 inches) to 200mm (8 inches) in thickness. The timber roof is of cruck construction, which in this case is sitting on earth-fast pad stones. This means that the roof load is transferred directly to the ground via the crucks.

Crucks are commonly built in to the walls in 'cruck slots'. While in masonry construction this is not particularly critical, in earth construction the thinning of the walls in these locations does introduce a slight weakness. If there is any movement of the roof structure or the foundation, it can often result in cracking around the cruck slot.

Each layer of earth is separated from the subsequent by a thin layer of straw. The straw is to provide protection during the drying process. Whether this layer is required in this situation is not clear, as the thin layers would tend to dry out much more quickly than a layer of 400mm (16 inches) to 500mm (20 inches) in height. Perhaps it is more important in this situation to prevent the walls from drying too quickly.

This technique is common to Dumfriesshire and right across the Solway Firth in north Carlisle. Adjacent to the field gate, the stone base course has been built up to protect the wall from being rubbed by the shoulders of passing cattle. Similarly, the doorways for the cart and the cattle sheds are protected with stonework.

Originally, the building had a turf and thatch roof. This has mostly gone, and the building is now under iron.

In almost every case, the earthen walls are built off a stone base course, which can vary in height depending upon local circumstances and the use of the building. The base course is usually of rubble stone, built using earth mortar and pointed using lime. Traditionally, most of these buildings would have been thatched using the local materials.

North elevation, Priorslyn

Interior crucks, Priorslyn

Close-up detail of earth lifts

Exterior east gable, showing
masonry wall protection
from cattle

Cottown after conservation works

Perthshire, Cottown Schoolhouse

This is a rich clay area, and is well known for the brickworks at Errol, Pitfour. It therefore makes sense that there is an earth-building tradition which pre-dates the brick in this area.

When used for building, the local subsoil is tempered with sand and coarse aggregate, and a large amount of vegetable material. The aggregate helps to condition the subsoil to form an almost stone-like mix. The aggregate also helps to reduce the amount of shrinkage, while the vegetable materials help bind the earth together and reduce the amount of cracking.

The vegetable material in mass wall construction serves three main functions: the straw acts as a binder, holding the earthen mix together and allowing easier application of the wet mix to the wall without it breaking up; it distributes shrinkage cracks throughout the wall, reducing the risk of serious cracking at key structural locations; and it assists with the drying of the mass wall.

Occasionally, vegetable materials are used to assist with the waterproofing of the walls or roofs. In some Mediterranean countries, seaweed is used to waterproof the earthen mix used for the flat roof.

Earth walls are normally protected with a lime render and lime wash. However, earthen finishing coats are common throughout the country.

At Cottown, the earth walls have been built in lifts of approximately 450mm (18 inches). No shuttering was used. The mix was kept fairly stiff and placed on the wall using a forked implement. (While there is considerable documentary evidence for rammed or shuttered earth techniques in Scotland, the writer is not aware of any extant historic examples.)

The earth would then be trodden down to compact it. The walls would be allowed to dry for a while with a layer of straw over it. Once it was fairly firm, the surplus earth would be pared down with a spade-type tool (similar to a targ spade) to form a finished face

West gable after conservation but before render application

both inside and out. The lift would be left to dry for a further period, dependent on the weather, volume of earthen material and type of mix. These techniques are all fairly labour-intensive, suggesting that this work was perhaps carried out by the community, with a good mud mason overseeing the works.

Changes to the original building can often be clearly identified when the render is removed. At Flatfield by Errol, an earthen wall supports an upper-floor wall constructed of stone.

The tradition in Angus is similar to Perthshire, as the geology of these areas is comparable. Logie

Schoolhouse in Angus is a property that has recently been repaired. The construction is mudwall, similar to Cottown. However, the walls in most cases were later clad in brickwork laid on edge.

The wall thickness was pared down to allow the brickwork to be built off the original stone base course. Again, like the blackhouse, whether this was to give the impression of a more permanent structure is not clear. However, in the village of Luthermuir, the writer was informed that even as late as the 1970s, home improvement grants were not available to earthen-constructed dwellings, and brick cladding

Logie schoolhouse after repair works

was a means of overcoming this hurdle. Virtually the entire village of Luthermuir was built of mudwall construction.

Throughout Scotland a great number of masonry walls have been built using an earthen mortar. As can be seen in Angus, the local sandstone is fairly soft and the earth makes an ideal mortar for construction. However, the primary reasons for using earth are its building characteristics and its ready availability.

In a great many cases the walls are then either lime-pointed, as is the case in Orkney, or lime-harled or washed, as is the case in Morlannich, Perthshire. Earth renders are also common and can still be found on internal surfaces that have not been exposed to the elements, as can been seen in the Grampian example (below).

LEFT: Earth mortar, Grampian

BELOW: Morlannich, Killin: lime-rendered

Close-up detail, clay and bool

Speybay

This area is very rich in glacial rounded boulder stones, particularly around the shoreline. This abundance of materials has been put to good use as a building material. The rounded stones present a challenge to the traditional builder, but by building the stone up on a bed of earthen mortar that is rich in vegetable material, the inherent difficulty of building with rounded stones is resolved. With careful planning and sound building skills, the layer of earth and straw between stones allows these stones or 'bools' to be put together to form herringbone and other decorative patterns.

Earth, perceived as a material used as a necessity born of poverty, should be looked on as a wonderful, eminently sustainable material. However, before embarking on a project using any form of earth construction, it is important to understand the characteristics of locally sourced materials.

Clay and bool cottage, Speybay

CHAPTER 5

CLAY

Bruce Walker

Scottish builders have used clay construction for thousands of years.

Despite being one of the earliest and most widely used building materials known to man, little academic research had been carried out into historic clay construction until recently. A lack of knowledge and understanding of this material by building professionals, together with its similarity in appearance and colour to the subsoil of any given site, has led to its presence often being overlooked in the study of vernacular buildings and in their repair. Lack of recognition has led to a misinterpretation of the nature of surviving clay structures, from prehistory to the twentieth century.

Clay construction can be referred to as 'puddled clay', and it should not be confused with earth or turf construction, both of which contain some form of vegetable fibre. The useful properties of clay as a waterproofing material or as a mortar were well known to early builders. The earliest surviving monuments in Scotland are chambered tombs. On the mainland, these were built in timber but were protected from damp penetration with a thick layer of puddled clay. In Orkney, where there was a scarcity of suitable timber, these monuments were erected using the local flagstone, which delaminates from the bedrock of the foreshore without the need for masonry tools. The importance of clay as a waterproofing layer was recognised when it was found that the chambered tombs which had been archaeologically excavated from ground level down to the masonry, then 'reinstated', were subject to perpetual drips from the vaults, even in dry weather, whilst those investigated via the original entrance tunnel were still dry after more than 5,000 years.

This suggests that the archaeologists excavating these structures failed to recognise the importance of the clay waterproofing layer immediately above the masonry.

Confirmation of the existence of this waterproof layer was made at Maes Howe, Orkney. This monument had been looted by Vikings, who broke into the chamber through the vault. Later in the twentieth century, the damaged vault was replaced by a modern vault, capped with concrete to protect the monument then covered with earth. Subsequently, the chamber suffered from dampness. In an investigation into the source of dampness, a test-pit was excavated on the edge of the concrete cap. This showed that water coming over the edge of the concrete cap had created a void, the top of which was level with the concrete, the bottom stopping against the 50mm thick clay layer protecting the original stonework. The clay layer did not erode in the same way as the earth fill.

Closer inspection of the interior of the Maes Howe tomb showed that the massive flagstones used as the floors and roofs of the actual burial chambers were laid on a fine clay mortar. The method of laying these massive slabs possibly involved pulling a slab into position over an earthen ramp, before excavating the ramp from under the slab, dropping it into position. The clay mortar may have been further plasticised by the addition of oil or fat, to allow the position of the stones to be adjusted to keep the line of the wall.

The situation is similar at the nearby Neolithic settlement of Skara Brae in Orkney, where the original archaeological investigations suggested that the houses had been dug into earlier midden material. A number of weaknesses are now recognised in this argument. Midden material normally accumulated on the surface of the ground, and was not deposited in a pit large enough to dig a settlement into. Also, midden material tends to vary in consistency and density. An alternative argument was put forward that midden material may have been used in a consolidated form as a fill between the flagstone faces to the walls, but that there had to be a waterproof layer under the floors and up to the walls to prevent water penetration from the surrounding sand. An

OPPOSITE: Clay has been used as a construction material since the building of some of our most ancient monuments: Maes Howe, Orkney

58

archaeological excavation was carried out to look for these features, with extraordinary results. Not only was there a layer of clay under the flagstone floor, but this continued up the outside of the outer flagstone face. The so-called midden material in the core of the wall had clay content, and the inner face of the flagstones had been finished in clay plaster, despite there being no clay in the immediate vicinity of the settlement. In addition to waterproofing, the Skara Brae houses were equipped with a number of small flagstone troughs sealed with clay at the joints. Those are thought to have been used to keep shellfish alive until required for cooking.

Without the examples of the Skara Brae masonry, these early uses of clay may have been lost. The properties of the material, however, continued to be utilised in construction, for instance, in the use

of clay joints in cisterns, lades, watercourses and as tanking, where the ground level was higher than the floor level in buildings. The slippery nature of hard-packed clay was also recognised, and many of the earliest stone buildings built off clay subsoil were protected from slipping by hammering pebbles into the surface of the subsoil, to create a friction layer between the subsoil and the base of the masonry wall.

The only form of clay construction to survive in abundance is the clay floor, which in early dwellings usually appears as a black mass. These floors were constructed in a number of ways, all of which involved some form of 'tamping'. After sheep become numerous in the Highlands, it was common practice to put as many sheep in the building as was possible and leave them overnight. Sheep's hooves made

Clay was used by Neolithic Orcadians in building both the structure and furniture of buildings: Skara Brae

an ideal tamping tool, as the weight of the sheep is carried on four small hooves, and the constant movement produces a very well tamped surface. Engineers recognised the potential of point loads in tamping, and most tamping tools were pointed or chisel-shaped on the impact area. Larger machines for consolidating the foundations for new roads, etc involved the 'sheep's foot' roller: a heavy roller with a series of blunt cones projecting from the drum.

Mortar used in Scottish masonry construction traditionally comprised a dryish clay loam which was 'knocked up': that is, worked and pounded until plastic, then used to bond the stones of the wall. The earliest buildings that used this technique were ecclesiastical ones in Scotland, Ireland and the northern counties of England. In limestone regions, lime was the preferred mortar.

By the fifteenth century, clay-mortared masonry towers were increasingly replacing the earlier timber and tempered-earth structures. Timber superstructures continued to be used in the towns, but from the thirteenth century these tended to stand on clay-mortared masonry, base courses or vaulted cellars and ground floors. In all these masonry structures, puddled clay was used to waterproof vulnerable areas, such as basements and wallheads.

Some of the finest examples of clay-mortared masonry were built around the turn of the nineteenth century to the twentieth century. Large houses such as Kinlochmoidart, Inverness-shire, are typical. At

Clay-mortared masonry towers increasingly came to replace those built of timber and earth: Dunstaffnage Castle

Kinlochmoidart the sandstone for the dressed portions of the structure, such as window and door surrounds, parapets and heraldic panels, was Dumfriesshire sandstone, whilst the rest of the external walls were of the local whinstone. The whinstone faces were pointed from the outset in cement, due to claims that this material was waterproof. The cement shrank on drying, leaving hairline cracks between the pointing and the stonework. This resulted in a build-up of moisture behind the pointing, eventually resulting in an infestation of dry rot. The building was dehumidified to protect the timberwork and moisture monitors were installed but, even after this treatment, when a section of cement pointing was removed, water ran from the opening for more than an hour. The only effective long-term solution to a problem like this is to remove the hard cement mortar pointing and to replace it with a traditional mortar, which will allow the building to 'breathe'.

Engineers were much more successful in their use of clay: aqueducts, canals, dams, docks, harbours, lades, piers, river embankments and sea-walls were backed with thick 100–400mm layers of puddle clay. The normal expression for this material was

62

Kinlochmoidart House, originally built with a clay core

Harbour walls were just one example of the many engineering structures in which clay was an integral construction material

'well-heeled', as it was tamped using the heels of a workman's boots to exert the maximum pressure. Nineteenth-century newspaper advertisements offered second-hand clay from the back of dock gates for sale. Engineering specifications must still exist for the type of clay required for a particular function.

One type of engineering structure in which the use of clay technology is generally misunderstood is bridges or viaducts. These are often considered by engineers as being constructed principally of masonry elements, but in fact they often have a rammed-earth core, protected by a clay layer sandwiched between the core and the road surfacing. Many bridges of this type were erected by General Wade in his road-building scheme designed to open up the Highlands to rapid troop deployment after the 1745 Rebellion. As the Wade bridges were built for infantry and cavalry, many had a distinctive hump in the centre, which became a problem when vehicles with long wheel-bases began to use these roads in the mid-twentieth century. The original method of construction had been forgotten, and many road engineers simply heightened the approach ramps and shaved the hump from the centre of the bridge, as was the case at Shira

64

Bridge, Argyll. This destroyed the centre of the clay waterproofing layer, resulting in water penetration into the rammed-earth core. The core gradually softened until it had no remaining structural capacity. The resulting pressure on the spandrels supporting the parapet caused the parapets to bulge, and in some cases collapse. Inappropriate and costly repairs included forming concrete casings within the masonry. The reinstatement of the rammed-earth core and waterproof clay layer would have been much more appropriate. Similar problems have been experienced on historic bridges with a flat carriageway, as tank manoeuvres during the Second World War and, more recently, four-by-four vehicle safaris, have dislodged the original hand-set road material. Railway viaducts were often protected in a similar way, such as Kinnaber Viaduct, Angus.

In the 1990s, dampness was found in some of the casemates in the walls of the mid-eighteenth century Fort George. The 'earth' fill was removed from the walls and the brick-built casemates were re-tanked. An excavator removed the soil from the top of the wall and exposed the top of the casemates. The soil structure was found to comprise 450mm of fertile loam, over 500mm of clay, over small shingle, in some instances going down to the original foundations, but normally varying between 0.5m and 3m in thickness. The original tanking was 150mm of clay well-heeled over the entire vaulted area. This was removed and replaced with asphalt. The shingle was replaced, but no attempt was made to reinstate the upper layer of clay between the soil and the shingle. The unfortunate result of these inappropriate repairs is that the surface-water drainage pipes at either end of each vault are now permanently wet. Despite the mistakes, the knowledge exists to identify the appropriate specification. The last major earth structure to be built in Scotland was probably the Megget dam, which provides Edinburgh with its drinking water and was constructed in the 1950s.

Scottish builders have used clay construction for thousands of years. Recognition of clay as a significant traditional building material will allow further research into its use and into the development of the skills required not only to identify and repair historic clay structures, but to develop its potential for modern construction.

CHAPTER 6

LIME

Roz Artis-Young

The use of lime covers every aspect of our traditional masonry construction.

Lime has been used as a traditional building material throughout the world, and in Scotland evidence of its use stretches back almost 2,000 years.

Scotland has a significant history of masonry construction. The earliest buildings were drystane structures, which gradually evolved into earth-cored walls. The Romans brought sophisticated masonry construction and lime technology to Scotland during their occupation of the British Isles, but it is probable that the use of lime in building had evolved prior to the Roman arrival, as it had in every other part of the world where limestone was an abundant material. Remnants of the Romans' great ingenuity in using building limes in a wide variety of applications, from aqueducts to baths and mass wall construction, can be found in Scotland, such as at the site of the Roman Baths at Bearsden, Glasgow. Later, the great castles and cathedrals employed building limes as a crucial component of their construction, using lime in all its guises as a foundation, construction and finishing material. Lime mortars developed not only as the core building medium, but also into sophisticated plaster finishes for internal and external use.

This can be seen in the many examples of Scottish fortified towers, built predominantly with rubble masonry, with dressed stone details at windows and doorways. External walls were often finished with the application of a harled or 'hurled' coat of lime mortar, trowelled across the rubble and finished flush with the dressed details, to enhance the weather protection and overall aesthetic appearance of buildings. Internally, walls were initially

OPPOSITE: A rural vernacular building harled and washed in lime to give a distinctive appearance

BELOW: Harling used on one of Scotland's most prestigious historic buildings: Stirling Castle

finished with relatively crude plaster surfaces, which evolved into flat, straight, squared walls ready to accept decorative finishes. The development of monastic establishments encouraged elaboration and detail in masonry work, utilising lime mortars as bedding and finishing materials.

Building practice and materials developed for high-status buildings gradually filtered down to the general population. The costs of extraction and processing for building mortars meant that lime was used sparingly in vernacular construction. Early domestic building construction used earth-based mortars, often reinforced with vegetable fibres like straw and turf: materials which were readily available and could be used without processing.

On the east coast, where there were many examples of clay-covered walls (stones bedded in a clay mortar), walls were finished internally and externally with lime as a pointing mortar or external coating. To make the material go further, lime was mixed with cheap local sources of clays for use in plasters.

During the seventeenth and eighteenth centuries, industrial techniques for the processing of lime were

70

Lime has been utilised effectively as a mortar in Scotland for many hundreds of years

developed. As production grew, so did the number of uses for the material in building, agriculture and industry. The use of lime, for instance, was found to 'sweeten the land' by rectifying acidic soil conditions, thereby increasing crop yields.

Economic industrial production reduced the price of lime and allowed its use in mortars for vernacular construction. The development of the traditional Scottish 'stane and lime' construction came with the Agricultural and Industrial revolutions. Many of the vernacular historic buildings which survive today are simple 'stane and lime' mass masonry walls, with roofs of local thatch, slate or pantiles.

The growing industrialisation of central Scotland in particular created high demand for agricultural lime – to ensure a consistent supply of foodstuffs for the growing population – and for lime-based building mortars for constructing stone tenements to house the rising numbers of town and city workers.

Certain limes noted for their setting properties were specifically used for foundations and for specialised applications such as lighthouses. A huge range of industrial applications, such as iron-making and other metal ore extractions, glassmaking, soap production and sanitation, relied upon lime.

The use of lime covers every aspect of traditional masonry construction throughout Scotland, and has shaped our culture and economy.

Building limes are used in construction in the form of mortars for bedding stone or brick masonry units; for plaster for finishing walls, internally and externally; as 'parging' to protect the joints on the underside of pantiles; or as limewashes over stonework or 'harl' (external roughcast plastering). Lime mortars are still used today to repair traditionally constructed masonry walls.

Lime mortars are made by combining lime with sands or other forms of aggregate. They are used

when still 'plastic' and, once in place, will set to form a relatively robust material. The durability of our traditional buildings generally demonstrates that mass masonry 'stane and lime' is an inherently sound method of building. Longevity is allied to ease of repair. A building may, for instance, need to be re-pointed from time to time, renewing mortar within masonry joints by removing weathered material and replacing it with new lime mortar. Coats of lime harling can be patched to match existing harling. A coating of attractive limewash can consolidate both extant remains of original finishes and new work. Likewise, internally, lime plasters can easily be repaired with the right materials and techniques.

Although lime mortars are tough, durable materials, they are flexible. Minor movement within a traditionally constructed wall is taken up by the relatively soft lime mortar. A hard, brittle material, such as modern cement-based mortars, would crack, allowing moisture to penetrate into a wall but trapping it behind the hard mortar and forcing it to travel into the stonework, ultimately causing deterioration of the masonry. The weatherproofing qualities of lime mortars are derived from their ability to aid the evaporation of moisture, allowing the building fabric to dry out. Many traditionally constructed buildings were finished with external coatings of lime harling or flatter rendered plastered finishes, which contributed to the effectiveness of the moisture holding and evaporation cycle, commonly

Nineteenth-century building pointed in lime

referred to as 'breathing wall technology'. In this way, lime mortars provide a degree of thermal insulation and control of condensation internally. Condensation could be controlled by the ability of internal lime plasters to absorb a certain amount of moisture and eventually distribute this both internally and externally through the connected pore structure in joints in the masonry wall, releasing moisture and maintaining internal humidity levels.

Masonry set in lime mortar can accommodate structural, seasonal and thermal movement without significant damage. Movement joints were unnecessary on traditional masonry walls, unlike modern masonry which requires the incorporation of regular movement joints in its construction, owing to the brittleness of modern mortar binders such as Ordinary Portland Cement. Any movement occurring in traditional masonry is compensated for by minute adjustment over many joints and beds. The mortar can yield more than the masonry, so cracking is not transmitted through the walls. Where small cracks in joints do occur, they may be resealed by the slow re-deposition of lime in solution. Lime mortars become soluble very gradually, so that minor ingress of water, causing localised solubility, results in migration through the solution from one area to another, where it re-precipitates in a form of self-healing.

Building using lime-based materials afforded the opportunity for adaptation. The soft, flexible mortar allowed valuable building stones to be prised apart, recovered and reused elsewhere. The use of lime mortars with their sacrificial nature, bringing with them the ability to reuse masonry, reflects our modern conservation philosophy of reversibility.

The aesthetic properties of building limes are unrivalled by most modern materials. It is the chameleon of traditional building materials and can be fashioned to emulate carved stone, fanciful decoration or vibrantly coloured lime harled and washed finishes, as well as resting harmoniously with the surrounding geology and landscape in which it is built. Unfortunately, with industrialisation and bulk production of cement-based products, regional textures and colours of lime harl, wash and mortars

are being lost. In the past, the texture and colour of masonry reflected its local source and surroundings.

Lime is the product of processing limestone by heating. The process is relatively simple and almost certainly derived incidentally from man's basic need for fire. Pieces of limestone used to ring the site of a fire would undoubtedly have been partially processed within the heat of the fire, leading to the discovery of heat as the means to transform limestone into lime as we understand it today.

Limestone is essentially any rock that contains a high proportion of calcium or magnesium carbonate, and it is the basic raw material for making building limes in Scotland. Limestone is a sedimentary material, made up predominantly of sea organisms laid down on the bed of the oceans. The layers were compacted, and where the sea floor met a land mass, the sea bed was forced under or upwards over the land mass, with the result that the limestone was outcropped as a sedimentary rock. Where it was pushed below the land mass, it was transformed by the enormous pressure and heat into metamorphic limestone, which can appear in other positions on the land mass because of the ever-changing geology. This slow process can take millions of years.

Scottish geology is complex. Some areas of Scotland were attached many millions of years ago to what we now call the Indian subcontinent, having moved with tectonic plate action from the equator regions to meet other sections of the British Isles, which were attached to the North American continent. The complex nature of the geology is reflected in the range of limestone types, as well as the incidence of limestones available to produce building lime. Central Scotland became the area of the country with the greatest industrialisation, for here the geology yielded abundant limestone, coal, fire clay and other clay types, in addition to iron ore-bearing rocks, all critical to the development of industry.

Geologically, limestone does not appear to be as abundant in Scotland as in other parts of the UK. However, this is simply a function of scale: a relatively small outcrop can yield thousands of tonnes of material. Supplies were, and still are, plentiful. Limestone was shipped in its raw, quarried state or as a processed material all around the coasts of Scotland, but it has never been used in any great quantity as a building stone, with the exception of the Keith area in Aberdeenshire.

Limestone deposits are to be found in all three main geographic divisions of Scotland: Carboniferous limestones (roughly 3 million years old) are found in southern Scotland and the Midland Valley extending up to the Highland Boundary Fault; Dalriadan limestones (roughly 550 million years old), often metamorphic (i.e., those compressed and heated geologically) in the Grampians, south-west Highlands as far as the Great Glen Fault; and both Cambrian and Jurassic limestones (roughly 500 and 140 million years old) are found in the north and north-west Highlands, with metamorphic limestones occurring in Shetland. Where limestone deposits were not available or too difficult to win, 'boat lime' (processed into quicklime) or cargoes of limestone were transported to coastal lime-burning facilities. Other sources of limes were derived from the shells of sea foods, such as oysters, which were processed and burnt for use in building.

Lime is manufactured by 'cooking' the limestone in a limekiln at a sufficient temperature to release the carbon dioxide content held within the stone. There are many types of limekiln, but field kilns can generally be found in rural areas, with the larger industrial kilns found at lime-burning complexes such as Charlestown, Fife. Old limekilns are often sited close to a limestone quarry, limiting the transportation of the stone.

Limekilns are usually masonry structures, built to contain both the raw limestone and the chosen source of fuel. Earlier 'clamp' kilns were simply holes in the ground filled with limestone and fuel, covered over with earth and set on fire until the limestone turned into quicklime. Coal measures of the Carboniferous Series predominant in central Scotland often occurred side by side with limestone, making this fuel the obvious choice. Elsewhere, where trees were in abundance, kilns were fuelled with timber.

The limekiln's structure and shape reflected the fuel used. Internal kiln pots linked to a wide firing chamber housed lengths of timber, while smaller

73

A lime kiln in the west of Scotland. Lime production was often a local affair.

Weathered lime harling acting as a protective coating to the stonework below: Stromness, Orkney

kilns reflected the size of coals, which were simply mixed amongst the limestone. Field kilns, often used for the production of both building and agricultural lime, were often 'flare' kilns, processing limestone in a 'one burn' operation to meet the needs of farmers and local building work. Industrialised limekilns utilised 'draw' kilns, constructed to allow a continuous feed of fuel and limestone to be delivered at the kilnhead, and continuously 'drawn' out as quicklime at the 'draw' holes.

The limeburner's job was to ensure the maximum conversion of limestone into useable lime with as little fuel input as possible. It was not an exact science, and the limeburner would simply look for a change in the flame colour to 'cherry red', indicating his charge had reached the appropriate temperature for the disassociation of carbon dioxide (roughly 850–900°C) and for yielding quicklime.

Scotland's last major lime producers ceased production in the mid twentieth century, but recently an experimental limekiln has been constructed at Charlestown, Fife, in an effort to research historic limeburning, and to re-establish limeburning for building in the twenty-first century.

The amount of lime used in construction depended on the geology of an area. Orkney, for instance, has no natural limestone, so clay was the predominant bedding material for the local sandstone. The finishing of surfaces relied on lime imported from limestone sources on the mainland and Shetland. The use of lime in Orkney was reserved for the finishing of joints and the surface of walls inside and out, using locally available materials to maximum effect. Materials were only imported when it was essential to improve the weathering capability of the building and, as wealth increased, the aesthetic appearance.

Although transportation of lime by boat had been going on for many thousands of years, the costs of transportation would have restricted the use of lime to those with wealth. Developments in transportation allowed greater quantities to be moved more quickly, making the material more economically viable. All but the poorest could eventually afford to use lime in greater quantities, and even in areas of Scotland with relatively scant lime sources, its use became common for virtually all buildings. With the exception of drystane and clay-cored walls (most of them finished with lime), the vast majority of Scottish buildings that predate 1920 are of 'stane and lime' or brick and lime construction, including tenements, town halls, barns, bothies, castles and cathedrals.

The strong link between masonry and lime has shaped the built environment of Scotland. The mortar between the masonry units, the surface finishes on the outside of our traditionally harled and rendered walls, the plaster on internal walls and ceilings, and the lime-harled colour of our townscapes all relied on lime. Lime has been used throughout Scotland; Even on St Kilda, the least accessible and most remote area of Scotland which was abandoned by its population in the 1930s, the hamlets are of 'stane and lime'.

Scotland may not possess large limestone deposits, but its complex geology made it possible to find outcrops of limestone across the country that were worked for farming and building for centuries. Even where no limestone was available locally, the proximity of the majority of the population to the sea, with its source of shells, made the use of lime widespread. Lime can therefore be seen to be not only an integral part of our built heritage but also vital to the maintenance and upkeep of all our historic and traditional buildings.

Lime mortar ready for use

CHAPTER 7

PLASTER

Gordon Urquhart

In the first two decades after 1603, there was a flowering of fine plaster decoration in castles, palaces and townhouses throughout Scotland.

Ancient origins of plaster

Plaster is one of the oldest manufactured building materials, with clay and mud plasters having been used to smooth and insulate wattle and masonry walls since prehistoric times.

The earliest documented use of plaster was in Mesopotamia around 9000 BC. Recent excavations in Jordan dating to 7500 BC have uncovered lime plaster used not only for walls and floors of buildings, but also for statuary. The most prolific plasterers of the ancient world, however, were the Egyptians, who burned limestone and gypsum to produce plasters that could take painted decoration on the walls of buildings and the surfaces of sarcophagi. By 500 BC, the ancient Greeks had perfected the use of a fine, white plaster made from burnt marble, using it extensively internally and externally, and often painting, staining or colouring it.

Knowledge of limeworking passed from Greece to Rome and eventually around the wider Roman Empire. Vitruvius, Caesar's military engineer, wrote extensively on plaster technology c.16 BC, the earliest detailed documentation for the craft. This knowledge percolated through France and England, and eventually to the empire's north-west periphery in Scotland. However, after the Romans departed from Scotland in the early third century AD, lime technology was in abeyance for hundreds of years. With Scotland on the far edge of Dark Ages Europe, the ability to produce lime mortar and plaster did not return to Scotland until later during the Christian period.

Lime production in Scotland

The traditional mortars and plasters used in Scotland were made from clay or lime (the binders), with aggregates (coarse or fine sand) and reinforcement (usually animal hair for lime products and chopped vegetable matter such as straw, heather or grass for clay). The manufacture of lime for mortar and plaster is based on a cyclical process in which the raw material is broken down into a malleable form that eventually cures and hardens into its finished state. Quarried limestone or shells (calcium carbonate) are 'burnt' in a kiln to drive off carbon dioxide and water content. The resultant 'quicklime' (calcium oxide) is then immersed in water or 'slaked' to produce soft lime putty (calcium hydroxide). The lime putty is mixed with sand to produce mortar or plaster, whereupon it cures by reabsorbing carbon dioxide from the atmosphere, thus returning to its original, hard calcium carbonate form.

The production of lime for mortar and plaster was always problematic in Scotland, for not only was the raw material relatively rare, but large amounts of fuel were also required for the lime-burning process. Scotland has never been a major producer of lime, unlike its neighbours England and Ireland, who were 'magnificently supplied with limestones' and to whom Scotland used to turn for imports.

Most lime-burning sites had small kilns serving a local area: a single farm, community or estate. Commercial sites had large constructions, sometimes with banks of contiguous kilns. Arden, south of Glasgow, had two pairs of kilns; Cambusbarron in Stirlingshire had 'an unusually long range of six limekilns'; and Charlestown, the largest site in Scotland, could boast a bank of fourteen kilns. Charlestown's operation exported widely along Scotland's east coast as well as into the Borders, occasionally to the west coast and as far north as Shetland.

Although the kilns varied in number, size and shape (round, oval, D-shaped or square in plan), the operating principle was always the same: fuel and raw lime were loaded from the open top of the kiln, whilst the large arched opening at the base drew in air to enable the firing process and permit extraction of the burnt lime. Lime-burning was a seasonal activity that ran usually from March to November, for not only did poor weather limit the effectiveness of the kilns, but the predominant product – agricultural lime – was not required during the winter months. Lumps of burnt limestone were removed from the draw-holes and either sold 'as is' or slaked on site to be made into lime powder for agricultural purposes or lime for building.

History of plasterwork in Scotland

In the early Middle Ages, stylistic and technological influences from Norman England, Ireland and the continent prompted the reintroduction of limeworking in Scotland. Evidence of this can be seen in the use of lime mortar in surviving structures from this period: for example, St Margaret's Chapel at Edinburgh Castle, Castle Sween in Argyll, the Church of St Rule at St Andrews, and the round towers of Brechin in Angus and Egilsay in Orkney. Although thin coats of lime harling were sometimes used on the exterior walls of rubblestone castles, lime plaster was rarely used internally at this time, as the formal rooms were usually clad with timber panelling or decorated with tapestries. Exceptions to this include certain ecclesiastical sites, such as Dryburgh Abbey by Melrose, which had plastered surfaces with painted decoration dating from at least the thirteenth century.

By the Renaissance period, wealthy Scots were decorating their interiors with exquisite painterwork applied to flat plastered walls and the timberwork of

walls and ceilings. By the early seventeenth century, highly decorated painted interiors were common in royal palaces, churches, castles and burgh houses from Dumfries to Kirkwall. Good surviving examples can be seen at Crathes Castle in Aberdeenshire, Traquair House near Peebles, and Holyrood Palace, John Knox's House and Gladstone's Land in Edinburgh. At this time, however, common vernacular buildings were still simple constructions made of readily available materials such as rubblestone or turf, with clay rather than lime used internally to provide insulation. This was the case for the majority of domestic buildings in rural Scotland until the end of the eighteenth century, and although town buildings might have been constructed with lime mortar, few would have been plastered internally with lime.

Early ornamental plasterwork

Soon after the Union of the Crowns, skilled plasterers from England and the continent brought the latest fashions in decorative plasterwork to Scotland, and

> with a verve and swagger alien to the scene they covered all before them . . . [and] in this creamy-white froth of plaster Scotland was provided with more exuberant animals, demi-gods, well-fashioned pendants and proud heraldry . . . than almost anywhere else.
>
> G. Beard, *Decorative Plasterwork in Great Britain*, 1975

In the first two decades after 1603, there was a flowering of fine plaster decoration in castles, palaces and townhouses throughout Scotland. The earliest surviving example of moulded lime plasterwork is

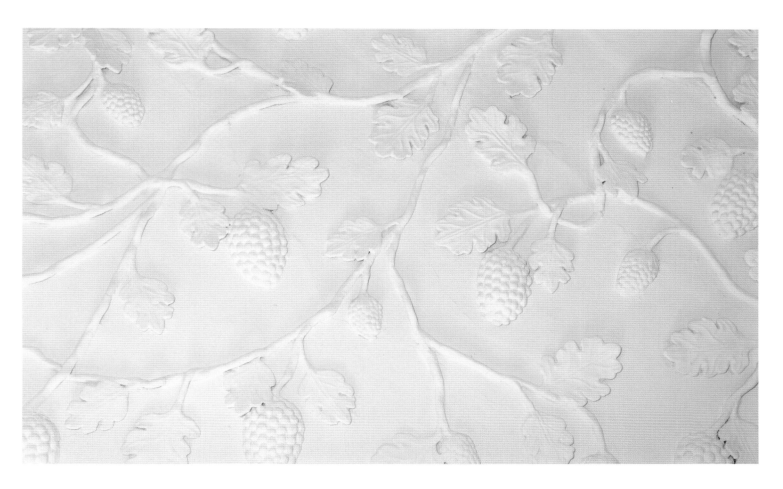

thought to be a fragment in the ruined Huntly Castle in Aberdeenshire, dated c.1602–8. Many important early examples, however, survive in their entirety: Pinkie House (1613) and Ballencreiff Castle (1617) in East Lothian, and Kellie Castle (1616) in Fife. Surviving documentation has shown that the same plasterers – from York – who were engaged at Kellie also produced the ceiling in King James' bedroom at Edinburgh Castle (1617).

In the 1620s, dramatic plaster interiors were created at the castles of Glamis (1620) in Angus, Thirlestane (1620) in Berwickshire, and at Muchalls (1624) and Craigievar (1626) in Aberdeenshire, as well as at Moray House (c.1628) in Edinburgh. Derivative of Elizabethan work in England, the ceilings were arranged in geometric patterns of moulded and decorated ribs, with modelled or cast ornaments inserted into the compartments between, often with motifs similar to those of older, painted ceilings, such as thistles, fleurs-de-lis, lions rampant and other heraldic emblems, biblical themes and simple floral patterns. The ribs often resembled traditional timber ceiling beams or fan vaulting and terminated with boss-like plaster pendants (notably at Craigievar, Winton House in East Lothian, and the Binns near Linlithgow). At Auchterhouse in Angus, the drawing-room ceiling boasted no fewer than twenty-four pendants.

A generation later, in the 1660s and 1670s, after the end of the English Civil War, and nominally to celebrate the Restoration of Charles II, there was a second wave of highly decorated plaster ceilings installed in great houses such as Thirlestane, Wemyss (Fife), Brodie (Morayshire), Arbuthnott (Kincardineshire) and several mansions in Edinburgh (Dalry, Gorgie, Merchiston, Stenhouse and Niddry). Kellie Castle, near Pittenweem, had a ceiling of elegantly modelled grapevines added around 1676, complementing its earlier plasterwork of 1616.

The exuberant, high-relief Scottish plasterwork

ABOVE: Floral patterns were often employed in seventeenth-century decorative plasterwork: Kellie Castle

OPPOSITE: Stunning plasterwork incorporating a wide range of mouldings, including the arms of the Union: Craigievar Castle

of the seventeenth century was not confined to the decoration of flat ceilings. For example, Moray House in Edinburgh's Canongate had elaborately decorated domed-shaped vaulted ceilings, whilst the great halls at Glamis, Muchalls and Craigievar castles had richly plastered barrel vaults complemented by extravagantly modelled Royal Arms flanked by caryatids in the overmantels. The stucco overmantel at Craigievar, according to HMSO publication *Behind the Façade*, has been called 'one of the most magnificent examples of British heraldry in this form'.

The most important new plasterwork in the last quarter of the seventeenth century was undertaken under the auspices of architect Sir William Bruce. At Holyrood Palace, Thirlestane and at his own houses of Balcaskie and Kinross in Fife, Bruce engaged the services of English plasterers John Houlbert and George Dunsterfield, as well as a Glasgow plasterer called Thomas Alborn. The ceilings at Balcaskie, noted in *Decorative Plasterwork in Great Britain* as 'among the most interesting in Scotland', include that of the Globe Room, where a double-combed ceiling terminates in a large belted globe pendant complete with signs of the zodiac.

The fine plaster decoration of seventeenth century Scotland was produced by combining several techniques. The flat areas were simply applied by trowels and floats, and moulded ribs and cornices were 'run' by a plasterer dragging a template along a fixed line.

> *In the earlier [seventeenth-century] work the ornament was modelled in the raw plaster by hand, with metal tools, or with the fingers, and by the freshness and quaintness of this work it is easily distinguished from the work of the subsequent periods, which was cast from moulds of wood or plaster.*
>
> G.P. Bankart, *The Art of the Plasterer*, 1908

The use of distinctive motifs permits the movements of itinerant plasterers to be traced as they worked their way around the country. For example, Glamis, Muchalls and Craigievar castles all have cast details made from the same moulds used in King James' 1606 palace in Bromley-by-Bow in London.

The eighteenth-century evolution

The heavily modelled, deeply enriched Scottish plasterwork of the 1600s evolved into a more academic, classical style in the early eighteenth century, when architect William Adam built or remodelled a number of great country houses, including Hopetoun near South Queensferry, Drum in Edinburgh, Mavisbank near Dalkeith, Arniston in Gorebridge, Yester near Gifford in East Lothian, Dun in Angus and Duff in Banff. Adam not only refashioned Scotland's most prominent country house, Hamilton Palace in Lanarkshire, but also built nearby the Duke of Hamilton's extraordinary banqueting house and hunting lodge, Chatelherault.

William Adam's buildings had magnificently modelled Baroque stuccowork, known to have been undertaken by English plasterers Samuel Calderwood (Mavisbank, Drum and The Whim, Peeblesshire) and Thomas Clayton (Holyrood Palace, House of Dun, Hamilton Palace and Chatelherault), and by Dutchman Joseph Enzer (Arniston, Yester, Dun and Duff). It is thought that Calderwood and Clayton often worked together, whilst Enzer had Scottish apprentices.

Baroque stuccowork in Scotland at this time was far removed from the rather naive and naturalistic subjects and motifs of the 1600s. Instead, a more rigid devotion to the Classicism of the Italian Renaissance developed, with heavier cornices, pediments, garlands, cartouches, urns, cupids and other Italianate ornament added to the plasterer's repertoire. Perhaps the most dramatic innovation of this period was the emergence of elaborate overmantels displaying

OPPOSITE: A wide range of tools and techniques were required by the craftsman to create Scotland's fine plasterwork

ABOVE: Baroque stucco work employing classical motifs

OPPOSITE: This time the classical influences are depicted in a plasterwork frieze incorporated into a wider scheme of plaster decoration

stuccoed trophies, or spoils of war, such as at Drum (Samuel Calderwood, 1727), the House of Dun (Joseph Enzer, c.1730) and Blair Castle in Perthshire (Thomas Clayton, 1751).

The impact of Robert Adam

Shortly after returning from his Grand Tour in the 1750s, Robert Adam emerged as the leading classical architect in Britain. Adam's unique style of Neo-classicism was delicate and refined, and far more intricate, linear and flat than that of his father's generation; low-relief stuccowork became the new paradigm. Adam's elaborate plaster ceilings became his main canvas, where his pure inventiveness acted as the unifying feature of his integrated room designs, linking with the wall decorations and even mirroring his designs for floor coverings. In buildings such as Culzean Castle and Mellerstain, Adam also revolutionised the use of colour in ornamental Scottish plasterwork, introducing coloured backgrounds of flat plaster to offset the white of the decorative details. Previously, ornamental plasterwork had been left in its natural off-white colour, or brightened with limewash.

Adam's plasterwork also incorporated a new material to eighteenth-century Britain: 'composition' plaster or 'compo', a patented material cast in moulds and fixed to a lime plaster background. Usually consisting of glue, resin and linseed oil, sometimes with pitch or whiting added (although there were various 'secret' recipes used by French, Italian and British artisans), compo was first developed in seventeenth-century France, and Adam was credited with introducing it to Britain around 1767 where it competed with recently introduced gypsum plaster (i.e., plaster of Paris) as the preferred material for precast applied ornament. Compo permitted thinner and more delicate ornamentation than gypsum and was praised in *Plastering, Plain and Decorative* for the 'sharpness and cleanliness' of the castings.

The use of composition and gypsum was not the only innovation in the plastering trade in eighteenth-century Scotland. Scagliola, an ancient formula for imitating polished marble made from lime plaster, glue, pigments and sometimes marble dust, became popular for finishing walls, stairs, floors, pilasters and columns. Papier maché, first developed in China 2,000 years ago, was used for architectural decoration in Europe in seventeenth-century France and imported to Britain in the early 1700s. Made from pulped paper pressed into moulds, it was primarily used for lightweight ornament, often in conjunction with fine wallpapers.

Later developments

A further revolution in traditional plasterwork came in the middle of the nineteenth century with the development of fibrous plaster decoration, where large sections of cast ornament are strengthened with a fabric backing and armatures of timber and wire. This new technology permitted the mass production of large, lightweight sections, thereby permitting a wide range of decorative features to be used in modest buildings for the first time. Cast in gelatine moulds and built up in thin layers of gypsum applied over a reinforcement of hessian or canvas, large pieces of fibrous plaster, reinforced by timber lath, were screwed in place. In large public buildings and other prestigious locations, single pieces of fibrous plaster could be as large as 4m².

The introduction of fibrous plaster coincided with the evolution in architectural style from the refined gentility of the Georgian and Regency periods towards the flamboyant exuberance of the High Victorian era, evident in buildings such as Glasgow's magnificent City Chambers (1888). The new reliance on the casting process, however, marked a 'divorce of the art from the craft', leading one Edwardian commentator, in *The Art of the Plasterer*, to bemoan that 'the plasterwork of the nineteenth century [is] so uninteresting, bad and uncouth'. Even the modern lime itself was called into question:

It is not seasoned – that is to say, the lime is not 'slaked' or toughened, or chopped up nearly as much as the lime which was used when most old cottages and houses were decorated . . . His sand also is vastly different, being riddled, and sieved of all the irregular qualities of the old ceilings which we admire so much.
G.P. Bankart, *The Art of the Plasterer*, 1908

This antipathy to technological advancement was fuelled, in part, by the emergence of the Arts and Crafts movement in the latter part of the nineteenth century. A few innovative architects, such as Robert Lorimer, ignored the soulless stylistic and technical developments of recent generations in favour of a revival of the handmade artistry of seventeenth-century Scottish plasterwork. Interestingly, Lorimer introduced handcrafted grapevine motifs adapted from his childhood home at Kellie Castle to several of his most notable creations: Rowallan Castle (1901) in Ayrshire, Ardkinglas House (1906) in Argyll, and Brackenburgh (1901) near Penrith in Cumbria.

Lorimer's approach was apparently driven by an emerging aversion to the stylistic excesses of the High Victorian period. When William Burrell commissioned Lorimer to remodel his house in Great Western Terrace in Glasgow, it is noted in *Lorimer and the Edinburgh Craft Designers* that he instructed the architect to 'chip away the gingerbread', preferring simple plaster covings. Similarly, when Charles Rennie Mackintosh remodelled his own Glasgow terraced house in 1906, he removed the cornices, preferring the flush corner junction of walls and ceilings as specified for Hill House in Helensburgh (1903) and Windyhill in Kilmacolm (1901). Although tastes were changing in the early twentieth century, not all Edwardian architects eschewed cornices

OPPOSITE: Fine decoration formed in a traditional plasterer's workshop

87

altogether; many simply used plain covings to cover the wall/ceiling junction.

By the early twentieth century, metal lath was beginning to supersede traditional riven timber laths. Metal lath was cheaper than the traditional timber lath, lighter and easier to install, did not warp and was resistant to fire, vermin and dry rot. In the mid-twentieth century, plasterboard (sheets of gypsum plaster encased in heavy paper) became the predominant means of cladding interior walls and ceilings, precipitating the steady decline of the art of the plasterer. In the past three decades, however, due to the renewed appreciation of Scotland's built heritage, there has been a resurgence in traditional plastering skills and a rekindled interest in the use of traditional lime. Unfortunately, there are no longer any active limekilns operating in Scotland; the last to close was located at Giffen, near Beith in Ayrshire, which ceased operations in 1972.

LEFT: Plaster can be used to form incredibly delicate decoration

OPPOSITE: More delicate, intricate decoration could be achieved as materials and techniques developed

89

CHAPTER 8

PAINT

Michael Pearce

These rare traces remind us that the fine stonework of prestigious medieval and early modern interiors was almost always painted.

Paint originated in prehistory and was first used in ritual body art and cave-painting. Now it is used for decorative purposes on all kinds of materials and contexts in the built environment, and also protects and extends the life of building materials. Roman houses and public buildings had painted decoration on plaster, both figurative scenes and framing elaborate architectural trompe l'oeil. Paint materials have been found in Celtic sites: orpiment, a bright yellow pigment, was found in a seventh-century context at Dunadd. The pigment was probably intended for use in manuscript painting and imported to Dalriada from the Mediterranean area. A thousand years later, in Renaissance Scotland, orpiment was mixed with the blue dye from woad to make the green used to paint trails of leaves on painted ceilings.

Traditional paintwork protected building materials by sealing them from the weather. Oil paints sealed wood, protecting it from moisture, which causes warping and strain and encourages rot. External ironwork needs a protective coating of paint, and sixteenth-century records show that the iron window grates of royal palaces were painted not only with red lead paint, but with expensive vermilion to enhance their colour. Harling coats and protects masonry, and limewash protects the harling. The colour of limewash was almost always improved by the addition of pigment. Limewash can be coloured by available local materials, either organic material or local ochres.

92

Although Scottish interiors can be distinctive and distinguished in regional type, the paint materials used do not show much regional variation. Although until the mid-nineteenth century almost all house painters made their own paints, they utilised the same refined materials available throughout the country. Highly decorated early modern Scottish interiors employed a highly specialised palette and realised Renaissance idiom in a distinctive manner.

Paint types

Paints are applied as a liquid and dry to form a durable film. The drying material is the paint medium; the colouring matter is the pigment. There are three main types of traditional paint: limewash, distemper and oil paint. Limewash is slaked builders' lime diluted with water. After application, the lime takes up carbon dioxide from the atmosphere, forming a hard crust of chalk crystals. Additional colour comes from added pigment. This could be available from local sources producing soft tints of red or yellow ochre. Limewash was used on harling and in modest domestic interiors, but distemper paints were more commonly used for inside work. Limewash has to be made on the spot, and the basic colour depends on the source of the lime. As well as pigments, green ferrous sulphide, known as copperas, could be added to limewash to produce an orange colour.

The medium of size or distemper paints was animal glue. This was produced from boiling the skins of animals. The size produced is a form of gelatine reduced from collagen. Some high-quality work required the use of glue made from boiling scraps of parchment. Royal accounts from 1538 mention the supply to painters of 'skrowis', which were pieces of waste kid leather from glove makers. The glue was diluted with water to form a gel, which had to be warmed to use as a paint medium. Pigments were

ground with water and mixed with the medium. The paint dried matte. The paint was used for plain walls, such as the interior of St Giles', painted white after the reformation, and for gaily painted Renaissance ceilings. Glue distempers continued to be used on plasterwork into the nineteenth century, because the matte finish was desirable. Oil paint became the usual finish for joinery.

Distemper paints do not keep very well and have to be made up on the spot by the painters. Colours used in interiors would not have been dictated by geographical variations but by vagaries of supply. The accounts for royal palaces in the sixteenth century show that pigments were bought in Edinburgh or St Andrews and sent on to painters working at Stirling, Falkland and at Hamilton Palace.

Unlike oil paints, distemper does not yellow, but it is not particularly hard-wearing and cannot be cleaned. Size paints were used in two different contexts: in prestigious rooms that would not receive much wear, or in service corridors that were often repainted. In 1807 the house painter Andrew Clason explained to his employer Robert Vans Agnew of Barnbarroch that size paints were used for the plaster walls of drawing rooms and bedrooms between surbass (skirting or dado) and cornice, or in passages. By the eighteenth century, ceilings and plaster walls were painted with size paints and joinery in oil, although oil paints were used on the plasterwork of high-status interiors.

Oil paint employed linseed oil as a medium. The oil is pressed from the seeds of flax. Although flax was cultivated in Scotland, linseed oil for painters may have been imported. Hope describes a visit to a Flanders oil mill in 1645. Local manufacture of linseed oil was promoted by the Commissioners for the Annexed Estates in the 1760s on the forfeited lands of the Earl of Perth. There had been a water-driven oil mill in the town since 1732, but this oil was probably destined for lamps. Oil made from walnuts was more expensive but was preferable in circumstances where linseed oil might turn yellow and darken delicate shades.

In the early modern period, oil paint was used for external work such as shutters, iron yetts, heraldry and statues. Internally it was used on carvings and woodwork, although distemper seems to have been preferred. Oil paint became much more common during the late seventeenth century when there was a fashion for pine panelling painted in imitation of walnut. The prestigious version was called 'prince's wood colour', and the slightly less elaborate and cheaper colours were known as 'walnutree' and 'wainscot'. These were used by James Alexander at Holyroodhouse during the residence of the Duke of York in 1684. While the exterior of windows at Panmure House in Angus were painted 'lead

colour' in 1671, a shade of bluish grey, the inside of the shutters was marbled. Only oil paint could be manipulated into the necessary swirling patterns and take layers of varnish to achieve a fashionable high gloss. In 1693 another painter, Walter Melville, was employed to match James Alexander's wainscot colour when a new suite of shorter hangings revealed the unpainted wall. He charged 10 shillings Scots for plain white oil paint per square ell, and 24 shillings to match the prince's wood colour. Although the taste for graining was short-lived in the eighteenth century, and panelling was repainted in plain shades, the fashion was revived in the nineteenth century.

Although Melville used plain white lead colour on new fixtures, and whitened ceilings and walls with size paints, he also cleaned and varnished old graining. At Edinburgh Castle he restored the decoration of the James VI Birth Room. His account preserved in the National Archives of Scotland refers to £36 Scots for 'painting King James his roume in the Castell of Edr in the fashione it wes before'. Paint was more often used to update and revive interiors and keep abreast of fashion. At this time, painters also strove to emulate the high gloss and decorative effects of wares imported from India and Japan. At Law's Close, Kirkcaldy, the home of a prosperous merchant, the panelling was grained and geometric patterns introduced in imitation of marquetry. These patterns have recently been reinstated after they were revealed under multiple layers of later plain painting. The rooms were restored by William Kay working for the Scottish Historic Buildings Trust. Thanks are due to the Trust and the charity Attention Fife, who occupy the restored building.

Plaster walls were also painted with oil paints: here there was a risk that the paints would react with fresh alkaline plaster and lose their strength. To mitigate this risk, interiors were either left unpainted for a time or initially finished with distemper. The distemper decoration could be washed off after a few years and a decoration in oils applied. Both William Leiper and Alexander Thomson recommended the use of 'ephemeral' stencil schemes in distemper in their buildings, prior to stencilling in oil; these schemes have been subsequently discovered at Lilybank House and Partick Burgh Hall.

Oil paints in the eighteenth and nineteenth century were in essence white lead paints with

OPPOSITE: Highly decorative marquetry pattern employed to impressive visual effect

95

96

Plaster mouldings were picked in and picked out in eighteenth-century practice

additional tinting pigments. Most pigments were commodities imported into Scotland. Some ochre pigments were native minerals. These are red or yellow iron ores which can be refined by roasting to produce a uniform red if desired. Yellow ochres were mined in Fife at Letham Glen and at Leven Collieries. Lead white was made in Newcastle, but there may have been manufacturers in Scotland. Although the copper mineral azurite is found in Scotland, the pigment was probably imported from Hungary or France. Other pigments were by-products of the cloth industry, based on dyes such as indigo and madder using the mordant alum. The cost of the raw pigment dictated the price and use of colour. Early accounts, such as the accounts of the masters of work for royal building and the Scottish exchequer records, give no detail of the origins of pigment. Apart from chalk and ochres, most pigments were imported in an international trade which has left little record. In

1538 Indigo, whatever its origin, was called Indigo of Baghdad, 'ind of badeas', and a red pigment 'rose of paris', but all that is known is that these were colours bought from merchants in Edinburgh and St Andrews. A deposit of graphite used for blackleading fireplaces was exploited on the Duke of Gordon's estate in 1721. In the nineteenth-century litharge, a form of lead oxide used as a drier was produced at the Duke of Buccleuch's mines at Wanlockhead, and lead white was produced from ore after 1889 at J.B. Hannay's works, Possilpark, Glasgow. Research has shown that during the sixteenth and seventeenth centuries, indigo was commonly used in interiors mixed with chalk to form a light blue, or with a yellow pigment to form greens used in decorative painting.

Even in the eighteenth century, most interior decoration made sparing use of shades apart from the 'common colours' made from earth pigments. These were white, stone colours, pearl, lead colour, cream, wainscot and chocolate. Of fifty sample cards with colours produced by Andrew Clason in 1807, which are held in the National Archives of Scotland, all but five of the fifty colours were shades of 'common colours' which cost twopence per lineal yard and could be used in distemper or oil. But even these colours may often have been purchased from London in the eighteenth century; in 1737, Lord Graham bought cream and chocolate-coloured paint for Buchanan Park there. In 1741 Sir Ludovic Grant bought paint for Rosedoe from a well-known London colourman, Joseph Emerton.

During the nineteenth-century, advances in inorganic chemistry made a greater range of colours available. Glasgow's chemical industry provided new pigments: J.J. White produced bright chrome yellows, reds and greens, which were used in the stencil pattern decoration made popular by Alexander Thomson and Daniel Cottier. From 1867 new azo-dyes were produced, heralding the ever-increasing availability of inexpensive pigments. Paints were also extended by barytes, a white mineral found near Aberfeldy, which bulks paints. Its use was developed by Orr's, a Glasgow manufacturer.

In the most ostentatious interiors, gilding was employed alongside paintwork. Leaves of finely beaten gold were made to adhere to carefully prepared surfaces with a varnish called an 'oil size'. Another type of gilding, where the gold leaf adheres to finely polished clay, called bole, was rare in architectural uses and only found on fixtures such as frames, or on details of heraldry. Gold paints were developed using finely ground brass powders. These 'bronze powders' became much cheaper after Henry Bessemer developed a new process for their manufacture in the 1840s. Bronze powder was strewn over a glaze of oil size, and varnished or mixed with a varnish medium.

Paints with very high gloss employing varnish media had been used for painting furniture in

imitation of Chinese lacquer since the seventeenth century. Similarly high-gloss paints were used on carriages in the eighteenth century, and these came to be employed on plaster and cast-iron interior features in the nineteenth century. A greenish and deep-gloss metallic finish achieved with layers of tinted varnish formed the basis of 'bronze painting', used for interior and exterior sculpture.

Painted drapery at Inchcolm Abbey, tomb of Bishop de Leycestre

Interior decoration

Most surviving medieval painted decoration is to be found in churches. Plain plaster was always covered with limewash or, more usually, size paint tinted with ochres. Decorative medieval schemes can be seen at the chapter house of Dryburgh Abbey, and where fragments of painting remain at Inchcolm Abbey and Torphichen Priory. Above the tomb of Good Sir James at Douglas, there are faint traces of painted tears, but the most striking example of medieval polychrome is the well-preserved painted fragments excavated at Glasgow Cathedral; in most cases, this medieval

Painted ceilings are important reminders of the richness of early interior decoration

work was covered over with plain white paint at the Reformation. The earliest surviving example of Scottish domestic painted interior decoration is a single stone vault rib found in the ditch of Dirleton Castle. Dating from the thirteenth century, it was decorated with a pattern of wide chevrons. Later painted decoration can still be seen on the corbels in the fifteenth-century hall of Craigmillar, which shows the remains of a pattern of a pair of trefoil leaves.

These rare traces remind us that the fine stonework of prestigious medieval and early modern interiors was almost always painted. Areas of plain or rubble masonry were plastered and painted. Wooden elements were also decorated, and generally it seems that within the homes of the elite the surfaces and nature of the building materials were always obscured. The keynote of the decoration of prestigious interiors was provided by textile hangings, tapestry and strips of brightly coloured silk brocade. Royals and nobles had a 'rooms of dais' centred around their cloth of

The most important sixteenth-century royal rooms had wooden ceilings, with applied ribs and carved insignia polychromed with oil paint by herald painters. In common with other late-medieval interiors in France and England, gold and azure were the chief colours. The profiles of rib mouldings were highlighted with red, white and blue. In lesser but still prestigious interiors, the joists and the underside of the boards belonging to the floors above were painted. These were painted with brightly coloured flowing patterns in distemper paints. Many of these ceilings have survived in part, concealed by later plasterwork, and some have been restored. Although most of the surviving or recorded examples of this decoration come from Edinburgh, Fife and the north-east, most interiors of this class throughout Scotland were probably similarly decorated during the time in which this style was popular (1580–1650). Gladstone's Land in Edinburgh has interiors with painted ceilings and accompanying mural decoration dating from around 1630. Painted ceilings and traces of similar mural decoration were found in several other properties in Edinburgh's Royal Mile, all of which bear a close resemblance to each other. Accompanying decoration on plaster and on stonework has rarely survived.

Wooden partition walls were also painted. Some survive at Northfield House, Preston, painted with similar patterns to the ceiling. At Stirling Castle, in 1629, the detailed painting account from an English painter who had settled in Glasgow, Valentine Jenkins, shows the use of grey and 'blew gray' in the lesser interiors of the upper floor of the palace. A plain wooden partition was painted in imitation of joined panelling with two shades of grey. At Kinneil House (near Bo'ness), the same painter drew imitation panelling on plastered walls. Joined oak panelling itself is rare in Scotland and it does not seem to have been as widely employed as in Tudor and Stuart England. Possibly the painted imitation

estate, with subsidiary hangings around the walls. The walls were plain-painted at the margins of the hangings, and highly decorative patterns were mostly confined to borders above the textiles. Such an interior survives in the north tower of Holyroodhouse, where a decorative frieze can be seen in the bed chamber. In the adjacent audience chamber, the same frieze survives behind the panelling. Beneath it, in 1625, parts of the wall were painted black to supplement black hangings installed at the death of James VI.

Argyll's Lodging, Stirling:
Corinthian splendour or flat
wood panels? Paint is employed
here to ensure that it is hard to
discern the difference

was preferred. A good section of seventeenth-century panelling at Rowallan Castle, with an internal porch and press, remained in situ until it was moved to a new castle designed by Sir Robert Lorimer.

Valentine Jenkins' Stirling account also shows that he 'helpit' – touched up – the sixteenth-century decorations in the royal chambers. His work at Stirling survives in part in the Chapel Royal. Painting in a similar style was visible at Amisfield Tower in Dumfriesshire.

Many of the later painted ceilings were executed on flat surfaces provided by tongue and groove boarding, the latest examples incorporating zodiac themes. Rooms lined with painted 'gribbed and sexed' boards, the old Scots term for tongue and groove boarding, survive at Culross Palace. During the seventeenth century, the Renaissance painted ceiling was superseded by decorative plasterwork. In some houses the unfashionable painted boarding was split up into lathes to provide a backing for new plaster ceilings and partitions. Today, decorative ceilings and plasterwork of the seventeenth century are usually painted white, although it is not known for certain if the ribs and moulding were originally picked out in colour. An imitation plaster ceiling painted by Valentine Jenkins at Kinneil shows ribs with a red background, which could indicate that moulded ribs were occasionally painted in the 1620s.

Remaining survivals of later seventeenth-century Scottish decoration provide surprises when compared with later Neoclassical treatments. One wall of the High Dining Room at Argyll's Lodging in Stirling consists of a cupboard with flat boards, which was painted with Corinthian columns by David McBeath in 1685. The pattern was continued around the plain plastered walls of the room. Grand rooms were more often completely finished with wooden panelling. Pine linings were painted, but some oak panelling may have been finished with oil. At Rowallan Castle, an early eighteenth-century room is panelled with pine frames, but the panels are plaster. Panelling at the turn of the seventeenth century was sometimes grained or marbled in surprising schemes bearing little resemblance to natural materials. Such patterns can be seen at Provost Skene's House and Midmar Castle, both in Aberdeenshire, and recently reinstated at Law's Close, Kirkcaldy. At Midmar and Provost Skenes, painters introduced into these swirling patterns little vignettes containing a nod to fashionable chinoiserie decoration.

A representation of a
Scottish vernacular interior
by Walter Geikie
© National Galleries of Scotland

The fashion for highly coloured and glossy faux finishes was short lived. Throughout the eighteenth century Neoclassical interiors were more often painted in pale and cream shades, and ornamental plasterwork was highlighted by 'picking in' or 'picking out' mouldings from tinted grounds. Skirting boards were sometimes painted in a chocolate colour throughout the century.

High colour was introduced by silk damask stretched from dado to cornice or, increasingly, block-printed wallpapers. Figurative painting was also introduced to Scottish interiors by artists such as James Norrie and William Delacour. Plaster panels from a house in Edinburgh's Brown Square, painted by Delacour in 1759, are on display at the National Museum of Scotland. These are landscapes in the manner of Francesco Zuccarelli. In the original scheme they were surrounded, like oil paintings, with wooden moulded frames and linked with floral motifs.

There is much less evidence to document the homes of ordinary working people. Interiors have not survived, nor would we expect any recorded evidence of their orignal appearance. In the early nineteenth century, a few artists made drawings of real interiors as preliminary sketches for genre scenes. The sketches of David Wilkie and Walter Geikie record cottage or farmhouse interiors. In these houses, decoration and comfort may have been secondary to the practicalities of domestic production. Interior stone walls were harled and painted with limewash or distemper possibly tinted with ochre. Joinery would have been painted with ordinary lead-based oil paints in the 'common colours'.

In the first half of the nineteenth century, David Ramsay Hay was a prolific decorator and self-publicist who produced six editions of his *Laws of Harmonious Colouring*. Apart from his complex colour theory of harmony and discord, Hay had definite ideas about the practice of painting, which were expressed in the later editions of his book. His work was most unusual in its employment of textured paint, which was manipulated to create a surface that emulated cloth. He patented a recipe for this paint in 1827. Hay's invention of textured paint did not catch on, although a less complex recipe was used to create the same effect at Edinburgh Castle's Great Hall in 1891. The whole wall surface was combed in two directions to give the impression of rich golden cloth with a red brocade pattern incorporating Scottish royal heraldry. Sadly, the thick paint failed, and the surface was repainted within five years.

Hay's use of stencil patterns were developed by a later generation of Glasgow painters and architects, of whom Alexander Thomson is the most prominent. Decorators such as Andrew Wells executed schemes devised by Thomson, William Leiper and Daniel Cottier. The Audsley brothers and Daniel Cottier exported elements of this Glasgow style to Australia and America. The designs of Charles Rennie Mackintosh incorporated Celtic influences, Art Nouveau, and Viennese Jugendstil into a distinctive Scottish product.

Stencil patterns in these Glasgow interiors were applied to lining paper. At Lilybank House, the plaster walls of the dining room were left with a slightly rough finish to help the adhesion of the paper, but above picture-rail level the finish was smooth and the frieze pattern painted directly on the walls. Alexander Thomson's interiors included the innovative use of pilasters made of varnished pine showing the actual figure of the wood. Stencil patterns on the walls were carried over the varnished wooden pilasters and architraves.

A further example of the decorative effects that can be formed using paint

103

CHAPTER 9

IRON

David Mitchell

Scotland was a world leader in the design and manufacture of ironwork for buildings . . .

The use of iron in traditional buildings is extensive, from simple ironmongery to structural and decorative elements, to complete iron buildings. In the nineteenth and twentieth centuries, Scotland was a world leader in the design and manufacture of ironwork for buildings, which was shipped across the world in large volumes to places such as South America, Australia, South Africa and India.

The manufacture of decorative and architectural ironwork made use of three types of ferrous metals: cast iron, wrought iron and, to a lesser extent, mild steel. All three were derived from ironstone, which is naturally occurring in various forms and may include other metal elements that give the iron differing metallurgical characteristics.

Ore was traditionally smelted in blast furnaces, using charcoal, coke, coal or peat. Lime was used as a 'flux'. The resulting iron was run from the base of the furnace, or 'tapped', and run into open indentations in the ground known as 'pig beds', resembling suckling pigs. A pig iron could be easily lifted by one man and re-melted in a small cupola furnace to make castings.

Castings were first made in open sand moulds in the ground (sometimes directly from the blast furnace), but as the industry developed, specialist moulding boxes and moulding sands began to be used. A mould of 'green sand' was prepared, so called because it was used in its raw or 'green' state. Clay particles amongst the quartz grains acted as a binder, becoming sticky with the addition of moisture. Fine coal dust was added to the sand, which burned out

106

ABOVE: Bonawe, like a number of Argyll sites, took advantage of local woodland as a fuel source

OPPOSITE: Ironwork in Central Station, Glasgow, by the city's Saracen Foundry

as the molten metal met it, helping to take the gases away from the casting and preventing gas bubbles from forming in the finished product.

A pattern (usually made of wood, but also of cast iron, lead or plaster) was placed on a board with a box around it. Facing sand with plumbago (graphite dust) was 'rammed' or pressed up against the pattern, with successive layers of sand built up on the top. The pattern would be carefully removed and the process repeated in the other half of the moulding box. Both sides of the box would be brought together and the molten cast iron taken from the cupola furnace and poured into pre-formed gates and risers. Once cooled, the box would be opened, the casting and gates removed, and the casting cleaned.

Most architectural cast ironwork used grey iron for manufacture. Cast irons have varying, but low, degrees of ductility. Impact resistance is minimal, although the material is excellent in compression and therefore ideal for columns.

Wrought or malleable iron was historically made in very small amounts, in forges developed as crude blast furnaces in eighth-century Spain. In 1784, Henry Cort invented the 'reverberatory' furnace, which allowed for consistent wrought-iron production on an industrial scale.

In the manufacture of wrought or 'puddled' iron, pig iron was 'boiled' within the furnace to a specific temperature and physical consistency by a highly skilled 'puddler'. The heated 'bloom' was removed at intervals from the furnace, worked beneath a hammer to drive out impurities, and then returned to the furnace, where the process was repeated. This repeated working gave the wrought-iron billets high iron content and a fibrous grain which provided excellent ductility and the ability to be worked easily. The wrought iron was classified in differing grades according to the quality and ductility of the bar. Wrought irons generally have excellent tensile properties and therefore work well when matched in

construction with the compressive qualities of cast iron.

Prior to 1855, steel was made by the 'cementation' process, where bars of high purity iron were impregnated with carbon by heating with charcoal. This was an expensive process producing relatively small amounts of material. The invention by Henry Bessemer in 1885 of the Bessemer Converter allowed steel to be produced on an industrial scale, although it was some decades before the material was readily available. Steel has low slag content and low carbon content. As a decorative architectural metal, steel had a minimal impact in Scotland until the late twentieth century, most smiths preferring to work in wrought iron, which is more malleable.

Decorative ironwork was largely undertaken in wrought iron until the latter half of the eighteenth century, when advances in the technology of manufacturing cast iron and increasing demand for a material that could be mass-produced led to a rise in the use of cast iron. The evolution of architectural cast ironwork in the nineteenth century had its stylistic roots firmly embedded in earlier wrought-iron forms, such as the significant wrought ironwork by smiths such as Jean Tijou in the seventeenth century.

Early uses of iron in an architectural context include hinges, strapwork, bolts, studs and locks. Throughout the fourteenth and fifteenth centuries, iron was increasingly used as an integral part of timber door construction. Door and window grilles, or 'yetts', constructed entirely of iron formed another layer of security behind a timber door, or acted as an external security measure, such as those found on the Palace block at Stirling Castle. The earliest reference to a wrought-iron yett in Scotland is in 1377, at Edinburgh Castle, when a reference was made in the Exchequer Rolls to iron purchased for the fabrication of a gate. The design and execution of the yett required significant levels of skill from the artisan fabricator. A secure structure was achieved by manufacturing intersecting 'L' shapes which were brought together in such a manner that they could not be taken apart once inserted into the masonry.

Stylistic developments in wrought ironwork advanced in the seventeenth century, leading to the design of features such as the gate overthrow, developed when the decorative cresting on gate panels became increasingly complex and cumbersome. It was not until the late seventeenth century that decorative ironwork was executed in Scotland on a grander scale. One of the earliest examples was the entrance stair to Drumlanrig Castle, Dumfriesshire dating from 1684, by blacksmiths William Baine and William Gairdner.

The fine wrought-iron balustrade of the main internal staircase of Newhailes, Musselburgh, built in 1685–6 by James Smith, is stylistically similar to the fine scrolled work found at Drumlanrig, with both appearing to have a French influence. The forged scrolls are particularly elongated and fine, in a style that was to be revived by Sir Robert Lorimer and his blacksmith Thomas Hadden into the twentieth century, during the Arts and Crafts revival in the use of wrought ironwork.

The ironwork gates and railings of the forecourt at Traquair House, Innerleithen, dating from 1698, incorporated wrought-iron flowers and the arms of the Stewarts of Traquair, which made a grand statement as visitors passed through the entrance gates by carriage. The design of the ironwork has been attributed to James Smith.

In the first decades of the eighteenth century, cast iron was increasingly used in component form within wrought-iron assemblies. The earliest use of cast iron in a significant context is generally considered to be the installation of cast-iron railings around St Paul's Cathedral, London in 1714. The comparatively low production costs of cast iron compared to the

labour-intensive costs of wrought-iron manufacture were pivotal in the increased use of the material in a decorative form. The evolution of architectural cast iron developed stylistically from earlier wrought-iron forms.

The use of iron in a structural context, to form the frames of industrial buildings using different types of ironwork in a variety of ways to maximise the benefits of their properties, was a significant breakthrough in the evolution of construction. Textile mills such as New Lanark, Deanston and Stanley employed the technology to maximum benefit in an attempt to provide for rapid and 'fireproof' construction.

ABOVE: Arts and Crafts ironwork by Thomas Hadden

OPPOSITE: Sixteenth-century iron yett at Stirling Palace

109

The early origins of iron production lay in the smelting of iron with charcoal, which developed in Scotland from the seventeenth century. Sir George Hay started iron production in Letterewe, Ross-shire in 1607, having had experience of iron production in Perthshire.

In England, the need for significant volumes of charcoal as fuel led to the destruction of forests, and restrictive measures were imposed to protect the timber resources. The search for other sources of timber that the English manufacturers could use began to have an impact on the iron trade in Scotland from the early eighteenth century. The relative abundance of timber in Scotland for the manufacture of charcoal, particularly in the Highlands, proved to be a temptation for many of the English ironmasters.

The threat of deforestation, as already experienced in England, was tackled vigorously by the Scottish parliament in 1609, when it published an Act Anent the Making of Yrne with Wode. The act restricted iron manufacture by threatening the confiscation of any iron produced in the Highlands, the principal area that had forests suitable for the production of charcoal.

The importance of the iron production and founding industry was recognised by the Scottish parliament as early as 1686, when it passed an act to encourage casting in Scotland:

His Majesty and Estates of Parliament, taking into consideration the great advantage that the nation may have by the trade of Founding, lately brought into the Kingdom by John Meikle, for casting of balls, cannons, and other such useful instruments, do, for encouragement to him, and others in the same trade, statute and ordain, that the same shall enjoy the benefit and priviledges of a manufacture in all points as the other manufactures newly erected are allowed to have by the laws and Acts of Parliament, and that for the space of nineteen years next following the date hereof.

Technological developments in the iron smelting industry allowed the Scottish industry to flourish in the late eighteenth and early nineteenth centuries. Significant discoveries by David Mushet and James Beaumont Neilson allowed the iron ore deposits of the central belt to be exploited. The backdrop of large-scale industrial development, and the emergence of a prolific engineering industry on the west coast of Scotland, provided the demand, skills and materials for the light castings industry, and particularly the architectural iron founders.

OPPOSITE: Structural ironwork, both decorative and functional on 'Greek' Thomson's Buckhead building

At the end of the eighteenth century, the Scottish iron industry was developing an infrastructure that was to form the basis of the growth of the architectural iron founding industry into the nineteenth century. Whilst a distinction should be drawn between the development of the production of pig iron, malleable wrought iron and iron founding, it is important to note that many of the early iron foundries were secondary businesses for iron smelters. The Carron Ironworks, for instance, smelted ore and used the resulting pig iron to cast goods or to manufacture wrought iron for a similar purpose, in addition to supplying cast and wrought iron to the marketplace. This was driven by a commercial realisation that manufactured goods could provide higher profits. Whilst other smelting operations, such as the Shotts Ironworks, founded in 1802, followed a similar route in operating a foundry, Carron was the company leading this approach in the eighteenth century and into the nineteenth century.

As the industry developed with the increasing demand for goods, in the early decades of the nineteenth century the operational separation of iron smelters and iron founders was seen as a logical step. The smelters had an increasing market at home and abroad to supply pig iron, while foundries increasingly specialised in engineering work, domestic goods or agricultural items. The use of the cupola furnace meant that pig iron could be re-melted at any location, with no requirement for a smelting works. Foundries could operate at a local level, but they could also position themselves close to national transportation networks such as canals.

The founding fathers of the architectural ironwork industry in Scotland were iron producers rather than iron founders. A relatively small number of key figures prompted and facilitated this development, but the significant degree of interaction between the key players had not been fully appreciated until recent research by the author mapped this out.

OPPOSITE: John Kibble's Palace of iron and glass, Glasgow

113

Scottish families that developed the iron and foundry companies were early industrial entrepreneurs in Scotland.

The Phoenix Foundry, established by Thomas Edington at the beginning of the nineteenth century, was one of the first major iron foundries in Glasgow with a significant architectural output. The ornamental gates of the Jew's Burying Ground at Glasgow Necropolis (1832), now lost, are believed to have come from the Phoenix Foundry, as are the gates of the Necropolis itself (1833). Yet it is to the Carron Ironworks that we must look for the principal development of cast iron as a decorative architectural medium.

Early castings, relatively plain and easier to mould and cast, began to be replaced by ornamental and stylised designs. Pattern-making and moulding skills developed, exemplified by the work of the Haworth Brothers at Carron. The Haworths were brought to Scotland from London by the architects James and Robert Adam, who became shareholders in the Carron Ironworks in the early 1770s.

The Carron Ironworks regularly commissioned the Adam architectural practice to design ornamental castings from 1759, with the family association lasting until the 1920s. The Adelphi Hotel in London has the earliest known example of architectural work produced by the company. It is likely that the first ornamental cast ironwork of any scale in Scotland was manufactured by the Carron Ironworks from 1764, when cast-iron railings were provided to the College of Glasgow.

The rise of the Adam brothers, the Carron Company and the Scottish iron founding trade were inextricably linked. Ironwork designs used by the Adams were executed at Carron in wrought iron, but often in conjunction with other metals such as copper and brass. Cast iron gradually appeared in their work towards the end of the eighteenth century, in details such as decorative finials employing classical motifs mounted on wrought-iron railings, such as those found around the tomb of James Bruce in Larbert Old Church (1786).

The bulk of the output from Carron was cast-iron water pipes, domestic ironwork and ordnance. The ironworks was to become famous for its domestic castings, with stoves and grates a speciality. The quality of detail and design was evident from the earliest output, and it was reviewed proudly some 200 years later in the company's trade literature.

The economic development of the Carron Ironworks into the early nineteenth century, however, was often precarious. Many other blast furnaces in Scotland were ceasing production. Carron's major advantage over its competitors was in the production of its own pig iron, which was turned into goods at source, together with its access to other markets by sea. As the first major foundry in Scotland, Carron provided a training ground for many of those who were to develop the industry in the future.

Scottish firms enjoyed success in the market for prefabricated buildings in the latter half of the nineteenth century. Whilst the credit for the introduction of cast-iron prefabricated facades currently lies with figures such as the American James Bogardus in the late 1840s, one Scottish firm was employing such techniques nearly twenty years before him. The Dundee Foundry Company made the transition from engineering castings to manufacturing a building in cast iron for Perth Waterworks in 1830, which may be amongst the earliest examples of cast-iron facade construction.

The benefits of iron buildings and elements for prefabrication were recognised by many Scottish manufacturers. In Edinburgh, the firm of William D. Young, subsequently William and Charles Young and Charles D. Young, was moving from supplying general wire fencing into prefabricated iron structures, mostly

OPPOSITE: Prefabricated cast-iron elements became used on many buildings, as here in Glasgow city centre

115

The splendour of the cast iron Ca' d'Oro, Glasgow

wrought but also in cast iron. This firm manufactured the iron structures for the Kensington Gore Museum and the Corio Villa in Australia.

Houses, churches, village halls and industrial buildings were created with iron frames, and were often clad in corrugated iron. The nature of the manufacture, creating a kit of manageable parts which could be easily assembled on the site of the building, combined with improving rail and sea links, made delivery feasible throughout Scotland. Corrugated iron as a roofing material, seen as requiring minimal maintenance, replaced thatch in many areas, creating a new vernacular Scottish style which is now valued on its own merits.

It was not until the 1830s that architectural iron-founding specialists were established in Scotland. The first of these was McDowall Steven & Co. of Milton Ironworks (founded 1834), followed by Walter MacFarlane & Co. of Saracen Foundry (founded 1850), George Smith & Co. Ltd of Sun Foundry (founded 1858) and Lion Foundry Co. (founded 1893). These five firms became famous throughout the world for their decorative cast ironwork, manufacturing not only spray fountains, clocks, canopies and shelters, but also theatres, railway stations and entire building facades.

Walter MacFarlane's Saracen Foundry became the world leader in the field. Although many of its projects overseas were recognised as exemplary – for example, the Madras Banking Hall, Mysore Palace, Sao Paulo railway station and many buildings in South Africa – only the Rothesay Winter Gardens and the Ca' d'Oro in Glasgow remain in Scotland.

Saracen drinking fountains are still found in Fraserburgh, Glasgow Green and Aberfeldy, with large spray fountains in Ayr (McDowell Steven Co.) and Paisley. The Paisley Gardens Fountain (George Smith & Co) is arguably the finest Scottish example.

At the end of the nineteenth century the overly ornamental styles of the Victorian era were replaced by softer forms. The Carron and Lion works in particular embraced the Art Nouveau style in the design of cast ironwork. The Arts and Crafts movement led to the resurgence of wrought-iron, hand-forged work, with Thomas Hadden becoming the leading exponent in Scotland to use the favoured natural motifs.

During the Second World War, a huge amount of cast iron was collected, as an act of civic duty, to stimulate the malleable iron industry, starting in Glasgow and extending rapidly to other towns and cities across the country. Post-war streetscapes were dramatically altered. Later, railings were sometimes replaced, but more often the cope stones remained, with just the stubs of the original ironwork or backed by the ubiquitous privet hedge. A small number of

towns, such as Stornoway and Lerwick, retain their pre-war ironwork and provide a unique and important glimpse of our earlier streetscapes.

The two world wars had a significant impact on the iron founder's skills base, as well as prompting a move away from ornamental casting to work in support of the war effort. The post-war housing boom utilised different construction techniques, and while the demand remained for rainwater goods and sanitary ware, as well as for constructional castings such as window mullions, the demand for ornamental work diminished sharply.

A handful of companies with the skills and knowledge to undertake traditional iron work now form the last remnants of the Scottish architectural ironworking industry. Cheap imports, the absence of training opportunities, and ever-tighter environmental controls now make it increasingly difficult for this sector to survive. This is borne out in the fact that there are currently no architectural iron founders in Glasgow, where once there were many.

Prefabrication using corrugated iron is not well appreciated

CHAPTER 10

LEAD

Roger Curtis

... simple to lay and maintain, and very durable, lead became the roofing material of choice ...

Lead has been a key element of traditional structures from antiquity onwards. It is an attractive, durable and workable material, and has historically had a wide range of uses in construction, from rainwater goods and gutters to flat roofs and cladding.

The earliest known examples of the use of lead in British building date from the Roman occupation. Lead pipes for the supply of water have been found in high-status Roman buildings. Sheet lead was used extensively as a roofing material from medieval times. Medieval monastic orders were active in lead mining and smelting. Archives record that Eadberht, Bishop of Lindisfarne, removed thatch roofing from a building, replacing it with lead, and clad walls in lead, in AD 635. Remains at the Carmelite Friary, Aberdeen, indicate that lead smelting and refining took place on this site from the fourteenth to the fifteenth century.

Prior to the Industrial Revolution, the use of lead was restricted to high-status buildings such as churches, mansion houses and public buildings. Relatively simple to lay and maintain, and very durable, lead became the roofing material of choice for the wealthier monastic and lay religious establishments. Labour-intensive methods of extraction, transportation and the significant level of skill required for its working, however, meant that it came with a high cost. Lead has always been valuable for reuse, and in the past, as now, it was often stolen from roofs. This, combined with repairs and replacement over the years, mean that examples of original lead remaining *in situ* are rare. Where old lead survives, scratched graffito

ABOVE: Jedburgh Abbey, showing lead-pitched roofs in use in the fourteenth century

OPPOSITE: A surviving example of an early style of leadwork, St John's Kirk, Perth

marks can often be seen, recording repairs carried out on the roof in the past.

A small number of examples of early leadwork in Scotland do still exist, such as the roofs of St John's Church, Perth, and King's College Chapel, Aberdeen, where there are records of an English plumber, John Burel, being employed to lead the roofs of the university in 1506. Fourteenth-century records describe John of Tweeddale being contracted to fit new gutters and leadwork to the choir of St Andrews Cathedral following fire damage, for which he was paid 25 merks. The lead lantern of the former St Ninian's manse, Leith, dating from 1670, survives in its original form, although the lead cladding is now mostly modern.

Nearly all the known examples of leaded spires in Scotland carried the lead cladding in short diagonal sections, to reduce longitudinal pull from thermal and gravitational pressure and to allow simple fixing. Each section was dressed and fixed down over a shaped softwood roll, a technique still used today. The spire at King's College, Aberdeen, was regularly repaired during the seventeenth century, where it was reported that the lead sheets were cast in individual sections, with the arms and date of the borough dressed into the sheet against a wooden mould. The

visual effect of the way in which the sheets were fixed to the spire gave a distinctive appearance, which was often recorded in early images of high-status buildings, as the details of the leaded spires of Old Aberdeen Cathedral on the Burgh Seal clearly show.

Roofs of a lower pitch further down a church building, such as on naves, aisles and transepts, used much longer sheets of lead, which ran down the building in parallel lines. Reports of church fires often refer poignantly to the molten lead running from the gargoyles, in a parody of their usual function in more fortunate times.

Like many historic trades, the provision of leadworking services and labour was tightly controlled by the borough guilds. Leadworkers in Scotland belonged to the Guild of Hammermen and the roof of their dedicated church, the Church of St Mary Magdalene, Edinburgh, appropriately retains some early leadwork. A master craftsman belonging to the Guild of Hammermen, such as John of Tweeddale, was a respected individual with a high status.

With so few examples of early leadwork remaining, little is known about the technical details of its use in the past for roof protection and rainwater disposal. It is likely that lead elements formed a decorative as well as a functional role. On grander buildings, decorated drain spouts were common. More lowly structures did not bother with rainwater disposal, although the French architect Violet Le Duc, writing in the nineteenth century, claimed that lead down-pipes were common in England in the fourteenth century. It is likely that they were not commonly used in Scotland until the onset of Classicism in the late seventeenth and early eighteenth centuries, where carefully designed and manufactured hopper heads and down-pipes took on an architectural role.

Lead was obtained in early times by the mining of lead ore in the form of lead sulphide, known as 'galena', which was recovered from shallow outcrops of ore near the surface. A good seam of galena could yield up to 60 per cent pure lead. Lead mining was carried out across Scotland, with known sites at Strontian, Glenlivet and in parts of Jedforest. The best known are the extensive remains at Wanlockhead. From that remote location in the mid-eighteenth century, 4,000 cart journeys were being made across the uplands to the Port of Leith sixty miles away, at a cost of 30 shillings per load. Lead from Wanlockhead supplied the Scottish market and was shipped to Europe.

Most early mining was carried out by hand, tunnelling into the side of the hill in a drift or adit shaft (where a tunnel was driven or excavated into the side of a hill at a shallow downwards angle), although by the late nineteenth century small explosive charges were used as part of the extraction process. After extraction, the ore was smelted and poured into 'pigs', the standard unit for the transport of raw lead. Each pig weighed about 60 kilograms. On reaching its destination, lead was cast on a base of sand in a tray up to 6 feet long and 6 feet wide. Cast lead tended to have a harder surface and was often characterised by a slightly uneven surface on the underside, caused by the sand base of the moulds. Until the nineteenth century, plumbers augmented their supplies by casting their own sheets from scraps, and joiners today still occasionally cast weights for sash windows themselves.

Milled or rolled lead was a later development, coming during industrialisation in the eighteenth century. Milled lead tended to be thinner and was thus more easily worked, but it had a higher risk

ABOVE: Aberdeen Burgh Seal again showing the diagonal fixing method on early leaded spires

OPPOSITE: Lead has been mined in Scotland for many centuries, often in remote areas such as here in Leadhills, Dumfriesshire
© Crown Copyright: RCAHMS

Lead has also been employed to roof larger areas, giving a distinctive feel to a structure

of cracking when exposed to significant thermal changes. Most modern lead sheet is rolled and is supplied in various thicknesses, referred to as 'codes', code 8 being the thickest, code 4 the thinnest. Widths can vary, with most sizes available up to 1 metre. Cast lead is, however, still available.

While lead is very durable when detailed correctly, it will not last forever. Most lead will last 100–120 years, depending on thickness and the type of loading. Decay of lead is caused by water erosion (particles in the water gradually eroding the thickness), thermal cracking (where the lead becomes brittle and cracks form, caused by the hardening induced by successive expansion in the heat and contraction on cooling), and point impact damage, often from falling slates or other debris. In some situations condensation can also cause decay problems. Repair is possible, by burning or 'wiping' on new sections, although eventually larger elements of a roof or gutter system will need to be replaced.

The adaptability of lead means that it is sometimes used in odd locations; for example, as a cushion between a series of load-bearing elements. The softness of the lead gradually allows the compressive pressure to become evenly distributed. Lead can be seen squeezed out from the base pad of the columns on the west end of St Paul's Cathedral, London.

Up to the present day, the use of lead in building has largely followed traditional techniques. Prior to the first use of high-temperature welding equipment in the early twentieth century, all lead had to be shaped or dressed by hand, using a simple range of tools, although in the nineteenth century some lead was joined by a technique called 'wiping', as mentioned earlier, where sheets or pipes were joined by partial melting of the adjacent surfaces, and dragging the molten material over the joint to create a bond.

Where an entire roof was covered in lead, the sheets were laid directly onto timber sarking boards. Modern practice often puts a felt material between the lead and the timber. To allow workable amounts of lead to be easily lifted by one man, and to reduce the risk of thermal cracking, lead roofs of nearly all types were formed in rows, joining the edges of adjacent sheets in rolls, mostly over a timber former but sometimes in a hollow roll. This technique gave the distinctive, lined appearance to cathedral and church lead roofs. Copper nails were used to fasten the edges of sheets, and exposed ends were 'dressed' or 'bossed' down. Roofs can take many different forms, from spires to flat roofs, but the leadworking techniques used in the past are very similar to those in use today, although fixing details vary slightly.

Lead was used to shed water from roofs via flashings and coverings, ridges, valleys and skew details, on slated or tiled roofs. Hips, ridges and their decorative variants have been used to strong architectural effect in a variety of situations, and the properties of the metal have to some extent influenced their shape. Fine embellishments were created, such as 'cresting', where small cast decorations were added to the ridge and hips of spires. Many ridges and apex flashings on turrets used the workability of lead to good effect, with ball decoration and other details as drops on a line of ridge. Pediments and cornices were often given a covering of lead to protect the stonework and to shed water.

Where water was concentrated, having been shed from a roof pitch, it was funnelled into a lead-lined timber gutter forming a series of trays, often behind a roof parapet. As lead sheets of only a certain length could be used, to avoid thermal cracking, the lead trays are separated by steps. Iron, or more recently copper, nails were used to fix the edges of the lead sheets onto the timber below. Where the lead ran against masonry or brick, the lead was recessed into a check or 'raggle' in the stonework and held there with small lead rolls called 'bats'.

Water was carried from the gutter via a lead pipe through the parapet into a lead hopper or 'pipe head', a square box at the head of the down-pipe. Made from thick lead sheet, hopper heads were often highly decorated and embellished with dates, initials or other patterns. Early down-pipes were usually square, socketed on the upper side for ease of connection, and often decorated at the sockets or 'faucets'. Water descended via a lead shoe into a drainage channel at the foot of the building. Sometimes the water was channelled into cisterns for storage, although few examples of these survive in Scotland.

While many of the tools and techniques of leadwork have remained the same for generations, the advent of small portable burning equipment in the twentieth century has altered the detailing of some elements of the more complex leadwork applications, and has allowed work to be finished quickly. Corner details are now folded and welded, rather than dressed in. The benefits of speed should be balanced with the additional risks arising from hot working on a roof structure, and the slightly different

125

appearance of the new details that lead burning gives.

Traditional leadworking skills are becoming scarce and the ability to produce high-quality leadwork has arguably become limited to a few contractors in the central belt of Scotland. Many roofing and slating companies have sufficient skills to carry out good work on the elements associated with roof slating junctions, but they leave the more complex leadwork to specialist plumbers.

The tools of the trade: lead sheet laid out for working with traditional tools

CHAPTER 11

GLASS

Robin Murdoch

Needless to say, glazing techniques varied with time and application . . .

Glass, as a building material in Scotland, has an interrupted history. First introduced in Roman times its use was limited since all the buildings liable to be glazed were military. Even then, it was only the most important or prestigious military buildings that used glass, such as bath houses, commandants' quarters, and so on. Places from that period where glass has been found include Inchtuthil and Ardoch, although it is likely that all of the permanent forts would have had some window glass.

A few shards of window glass have been recovered from immediate post-Roman contexts in Scotland but it is not until the great cathedral building phase of the twelfth century onwards that glass appears again in any quantity. For a considerable time afterwards it seems that it was still restricted to ecclesiastical use.

Glass is made primarily from three basic raw materials: a source of silica, a fluxing alkali and a stabiliser. The first is normally derived from sand and would vitrify on its own if the temperature was high enough. Unfortunately a high temperature would need to be maintained in order for the glass to be worked. Additionally, the costs and consumption levels of fuel to attain and maintain these temperatures would be prohibitive. To overcome this a fluxing alkali is added, either sodium-based or potassium-based, which drastically reduces vitrification and working temperatures. However, the resulting compound, often called water glass, is actually soluble in water, and a stabiliser in the form of lime is added. The durability of glass varies widely, soda glasses being

OPPOSITE: Small pane sizes are in evidence here, in what is believed to be the oldest in situ coloured glass in Scotland: Magdalen Chapel, Cowgate, Edinburgh

generally more resistant to breakdown than potash, although other constituents in the mix can affect this. Suffice to say that the constituents used to make glass affect everything from durability and colour to optical quality and hardness.

Sourcing the raw materials for glass making was relatively straightforward as silica sand of suitable quality, and lime for stabiliser, occur in plentiful quantities in Scotland. The same is the case for the sources of alkali for fluxing and the refractory clay for making the crucibles or 'pots'.

Generally speaking, most glass made without a deliberate colourant will have a green tinge. This is due to the presence of iron compounds in the raw materials or the clay of the pots. It was a long time before the glassmakers understood that the solution to producing colourless glass was to adjust the chromatic balance rather than attempt to purify the raw materials. Manganese was found to be a good decolourant; its addition to the batch should theoretically produce a purple tinge but, if the quantities are right, this neutralises the green induced by the iron leaving a colourless metal (glass in its unworked state).

The quality and durability of the clay to make the pots was extremely important. Failure of the pot could result in the loss of valuable batches of glass, and even in damage to the furnace, resulting in considerable financial losses. Clay, suitable for making pots, was available in several parts of Scotland – Fife, for example. However, records show that it was also imported from England, and even France and Germany in the early seventeenth century during the formative years of the Scottish industry. *The Scottish Glass Industry 1610–1750* notes that John Ray reported seeing glass being made on the shores of the Forth in 1661 and remarked that the pots were made of tobacco-pipe clay. Small fragments of broken pot were often used as 'grog' or temper in the manufacture of new ones, which helped to reduce cracking.

The fluxing alkali was also carefully chosen. Throughout the medieval and immediate post-medieval periods, the predominant fluxing alkali in northern Europe was potash derived from forest trees and plants. Southern Europe tended to use soda-rich barilla, a group of plants growing on the Spanish littoral.

Late in the seventeenth century, British manufacturers started to use kelp as the source of alkali. This introduced more soda into the melt and improved durability. Kelp remained a major alkali source until the early 1820s when Le Blanc devised his method of making sal alkali in a chemical process. This quickly became the base alkali for soap and glass manufacture.

The reason why potash was the predominant fluxing alkali in the medieval period was availability. When the Moors took control in southern Europe in the tenth century, the supply of natron, a naturally occurring soda alkali commonly used as a flux in glass manufacture, was cut off. Northern European glassmakers turned to potash instead. This was derived from the burning of forest plants, with trees, ferns and beech being typical. For reasons that are still not fully understood, potash fluxed glass can deteriorate more readily than soda, in many cases to an extent that it is not recognisable. This could be giving an unreliable indication of how much actually existed. It seems most unlikely that the owner of a large house or castle, having seen glazing in churches, would not aspire to have this material, which would admit the light and exclude the elements, in his own property.

Notwithstanding, it is generally accepted that glazing does not begin to appear in secular structures in Scotland until the fifteenth century. Perhaps because the technique derived from ecclesiastical glazing, quarry or pane size still tended to be small

Early crown glass showing characteristic curving striations

and was constructed in a framework of lead cames, these in turn being supported within iron frames or by iron glazing bars. The large pane sizes that we see today appear not to have existed in the early days, and the ability to see the outside world would have been somewhat restricted.

The sixteenth century saw a gradual move away from what were mainly defensive structures with small window apertures towards more user-friendly architecture with far greater areas of glazing. This trend would reach its height in the eighteenth and nineteenth centuries with the building of most of the great country houses.

Up until the nineteenth century there were only three ways of making window glass: broad (cylinder or muff), crown (spun) and plate (cast).

Broad glass was started in the same manner as if blowing a bottle but with further blowing and

swinging of the paraison (the initial blown globular shape) over a pit a long glass cylinder was created. This was reheated, the ends were cut off, the side was slit and the glass was flattened onto a surface to create a rectangular pane.

Crown glass started off similarly, except that at an early stage a solid rod (pontil) was attached to the closed end of the paraison and the blowpipe wetted off. After reheating, the blowing aperture was opened out and then the pontil was spun quickly. Under centrifugal force the glass flashed out into a thin disc anything up to 5 feet (1.5m) in diameter. The pontil would continue to be spun until the glass was stiff enough not to sag. Glass is a supercooled liquid, not a solid, and it stiffens as it cools.

Both broad and crown glass would be annealed, that is, cooled slowly in a lehr, a special furnace for that purpose. This was to release internal stresses

created during formation. Failure to anneal could result in the finished product being more prone to cracking or shattering under mechanical or thermal shock.

Although plate glass was made at one time by blowing thick cylinders, from the late seventeenth century onwards molten glass was simply poured onto a lipped table. However, this process was not in general use until the mid-nineteenth century other than for the production of mirror glass. It was heavy, expensive and required much grinding and polishing. By the later nineteenth century a rolling process had been introduced which also allowed the texturing of one or both surfaces. Textured plate glass was, and still is, used to provide privacy while admitting light. Plain plate glass is, of course, most commonly seen in shop windows where a large clear area of glazing is desirable.

Pane sizes in broad glass could be considerable, but optical quality was poor. The fact that the glass came into contact with some sort of surface during flattening meant that one side almost always had blemishes from that contact and lacked fire polish. Crown glass, on the other hand, was 'finished' in clear air and often had excellent optical qualities depending on the skill of the glassworker. Crown glass often has curved striations where the rate of flashing was not consistent. The disadvantage of crown glass was in pane size since there was a limit to the size of individual panes that could be cut from the disc. By the late eighteenth century, crown discs or tables could be made 5 feet (1.5m) in diameter, and this seems to have been the upper limit. Gosford House has crown panes of 28 inches (71cm) high. However, slightly smaller tables appear to have been more typical. Surviving cutting details

A contemporary eighteenth-century depiction of the art of manufacturing broad glass

from the early nineteenth century indicate that 17.5in × 11.5in (44.5cm × 29.2cm) was about the maximum that could be cut from a 49-in (125cm) table, and even then only two per table. The thickness towards the centre, or 'bullseye', of a crown disc varies rapidly and this would have given problems in mounting. Crown discs were also frequently slightly bowed, which would further complicate mounting.

Contrary to what most Christmas card designers would have us believe, the 'bullseyes' were seldom used for glazing and certainly not in prominent positions. The limited size of crown panes probably had a direct bearing on the development of multi-pane sash windows, where matrixes of six, nine or twelve were subdivided by wooden astragals.

Many modern structures have mock crown centres as features but these are generally pressed from panes of uniform thickness and bear only a superficial resemblance to the originals.

While both codes, broad and crown, were used for window glass throughout the medieval and post-medieval periods, it rather depended on where the glass was made. Broad glass was favoured by the Lorraine makers, and crown by those from Normandy.

Broad glass continued to be made in small quantities, but the use of crown began expanding about 1680 and was by far the most popular from the middle of the eighteenth century to the middle of the nineteenth century. In Scotland the great crown works at Dumbarton were founded about 1771, and by 1816 they were producing the equivalent of one-third of all English output. However, by the 1830s the company was in trouble, output became intermittent and the whole works shut down around 1850. Financial matters were part of the problem, but technological changes in the industry were probably more significant.

In the early 1830s an economical method of polishing broad glass, perfected in Germany, was introduced by Chance Brothers in Birmingham. Cost savings in production and labour were considerable, and the required skills levels were lower. This new 'sheet' glass quickly usurped crown as the preferred form of glazing. Crown production did continue for a small and specialised market, but its commercial days were over. Pilkington's in St Helens ceased crown production in 1872.

Needless to say, glazing techniques varied with time and application. The normal arrangement in medieval times was small, sometimes irregular, quarries or panes held in a matrix of lead cames. Typically H-section, the cames were made with opposing narrow channels into which the edges of the panes would be slotted. The ends of the cames would be soldered together to build up the matrix.

In the great medieval cathedrals and churches there were often very large window apertures. These would be subdivided into smaller areas by masonry mullions and transoms. Further subdivision took place by the insertion of wrought-iron frames or glazing bars into which the leaded matrices would be fixed. There is a limit to the practical size of an unsupported lead matrix.

A significant percentage of medieval ecclesiastical windows appear to have been decorated; of course, many of these decorations were on religious themes. Glass decoration took two basic forms. The first was coloured or stained glass, where a colouring chemical was added to impart a particular colour tint. This could be done at the raw material stage or by adding the colourant to one face of a finished clear glass pane and firing to 'fix'. Stained glass was commonly found in figurative window designs.

The second form was painted glass, where clear (in reality, usually slightly greenish) glass was surface-painted on one side with red oxide paint and the quarries reheated to 'fix' the pattern. Repetitive foliate designs and areas of cross-hatching, the latter

OPPOSITE: Characteristic distortion often found in pre-mid-nineteenth century glazing

RIGHT: Decorative glass in a secular context, still with small pane size: Great Hall, Stirling Castle

OPPOSITE: Ecclesiastical buildings were amongst the first to use glass on a significant scale

in the early period, were particularly common, and the painted side was normally internal to the building. Red oxide painted glass is usually termed 'grisaille'. It is believed that some of the more austere religious houses eschewed colour and only had grisaille-decorated glazing, if any decoration at all. Cistercian houses, of which Melrose Abbey was the earliest, fell into that category.

Not surprisingly, much early secular glazing appears to have been influenced by church techniques. Decorative panels with religious or secular themes were often found in large houses. These were highly prized and frequently moved house with the owners. To facilitate this, the decorative central panel was often surrounded by a narrow border of plain or lightly decorated glass. This border was simply sacrificed to allow safe removal of the important central panel.

In Scotland up to the later seventeenth century, virtually all glazing appears to have been fixed, that is, the windows could not be opened to admit fresh air. The solution to this was to glaze only the upper part of the window aperture and place two hinged wooden shutters in the lower part. A modern restoration of this technique can be seen at Culross Palace in Fife.

It seems likely that the Reformation encouraged changes in glazing content to an altogether more austere style. This typically consisted of a matrix of small 'lozenges', diamond-shaped panes, but still mounted in lead cames. This style would remain common in secular buildings well into the eighteenth century and in church glazing throughout the nineteenth.

Sash and case windows were formed in two parts, with panes of glass supported by timber astragals

Jill Turnbull's *Scottish Glass Industry* records that a visitor to Edinburgh in 1636 noted the lack of glass windows, observing that 'few or none are discerned towards the street'. However, Thomas Morer commented in 1686 that the new houses are 'made of stone, with good windows modishly framed and glazed'.

Fixed upper glazing with lower shutters continued well into the eighteenth century and, remarkably, oiled paper was still being used as late as 1732. However, towards the end of the seventeenth century a radical new style had begun to appear: moveable sash and case windows. These consisted generally of two frames or sashes, one or both of which were moveable, set into an outer case. The sashes could be moved by sliding up or down, and many were counterbalanced with weights inside the outer frame to stabilise them at whatever aperture was desired. Those without counterweights would be wedged, pinned or dowelled as required. The earliest versions appear to have had a fixed upper sash and a moveable lower, although later it was normal for both to move.

Although not exclusively, the introduction of sash and case windows heralded a move away from small, leaded diamond-shaped quarries to slightly larger rectangular panes held in position by astragals, vertical and horizontal wooden subdividers. Windows were multi-paned, and the earlier astragals were of substantial cross-section. This type of glazing remained very popular until the mid twentieth century.

The typically small size of the panes was because of two factors: cost, and the type of glass used. As has already been noted, from

Broad glass showing
characteristic bubbles
(seed) and blemishes

about 1680 crown or spun glass became the most frequently used throughout Britain. Because of the limited size of a crown disc or 'table', it follows that the maximum useful pane size would likewise be limited. Smaller panes also meant more could be cut from the table, reducing unit cost even further.

Of course, much larger panes could be obtained if they were broad glass (usually from France), but this was typically rather green-tinted, full of seed (small gas bubbles) and surface-blemished from the flattening process. Polished varieties could be obtained at extra cost, as could plate glass, which was even costlier since it had to be polished on both sides in a relatively long process.

The Palladian revival in Scottish architecture in the early eighteenth century saw window sizes increase – in some cases to very large areas, with heavy astragals remaining the norm for mechanical strength. By mid-century, however, Palladian had given way to Classical, and while window sizes remained large, pane sizes increased and the number and cross-section of astragals reduced. There were interesting variations on the sash theme: the small single-sash attic windows in Gunsgreen House, Eyemouth, c.1754, slide up into a recess in the wall head.

One significant difference in terms of the installation of sash and case windows in Scotland and England was that in Scotland they were almost always recessed in an ingo (where stonework turns at right angles to the main external face of a wall towards the window), whereas in England they were invariably fitted flush with the outer wall.

In the late eighteenth century and nineteenth century, sash and case windows appeared in many variations, especially in the grander houses as the architects of the day adapted them to their particular designs. The eighteen-pane window (two sashes of nine) of the first half of the eighteenth century gave way to typically twelve panes (two sashes of six) in the latter half. This was achieved not by a reduction in window size, but by an increase in pane size.

Round and pointed arch windows became more common, as did the use of wrought iron in astragals to provide the necessary strength in thinner sections. Sash doors were also made, and in many cases fenestration was designed to reach floor level to improve light admittance and improve the view.

A major stumbling block to the use of glass in windows from the late seventeenth century to the mid-nineteenth was taxation. Excise duty on glass and a tax on windows (that is, the total number of windows in a dwelling) was introduced in 1695 to fund the wars of William III. The excise duty was dropped in 1699, but the window tax was not. Excise duty on glass was reintroduced in 1746 and it, plus the window tax, were substantially increased throughout the eighteenth century.

Minor relief came in 1823 when window tax was halved, and again two years later when the number of 'free' windows was increased from six to seven. Excise duty on glass was finally dropped in April 1845, and window tax ended in April 1851.

While this should have been good news for the indigenous industry, the situation was not quite so straightforward. The levels of excise duty had kept cheap foreign imports (Belgian, for instance) very close to British-made prices, but the removal of taxes revealed a considerable differential in bare production costs. Crown glass had benefited from excise duty because it was typically thinner, hence lighter, than broad or sheet. Removal of excise duty no doubt contributed to the demise of crown glass, albeit in conjunction with improvements in sheet-glass production.

During the century or so of glass and window taxation, many structures were built with blind windows for economy. Recesses were created in building facades where windows would normally

Early sash and case windows incorporated small panes of glass: here in John Knox House, Edinburgh, twenty-four separate panes are used

Many intricate glazing patterns
came to be used in the nineteenth
century, here in the creation of a fan
light above a door

have been placed, in order to maintain architectural symmetry. These recesses were sometimes painted to simulate windows. In some cases, the windows were actually installed and glazed, but then blackened with paint or fabric in the interior.

While the sash window remained the principal style of fenestration into the early nineteenth century, French influence in particular was seen in inward-opening side-hung casements. A mix of sash and casement windows can be found in many early to mid-nineteenth century Scottish mansions. The nineteenth century also saw a throwback to an earlier style in the reintroduction of stone mullions and transoms.

In common with many other industrial processes, manufacturing techniques for window glass underwent considerable change from the late nineteenth century onwards. Improved sheet, introduced in the 1830s, still required hand-blown cylinders, and more automatic methods were continuously sought. Experiments with drawn sheet in the 1890s failed, but drawn cylinder proved practical. However, this process only flourished from 1920 until around 1930, at which time drawn sheet was successfully introduced. The most recent major development was Pilkington's float glass, made by literally floating the glass on a bed of molten tin. This method was introduced in 1959.

Other nineteenth-century innovations included etched glass, where designs could be formed by etching with hydrofluoric acid, wax being used to mask the rest of the surface. Sand-blasting was also introduced at this time, where a jet of sand particles produced an effect similar to etching. Etched and sand-blasted glass were used both internally and externally for windows, and frequently for mirrors –

those in public and licensed premises being typical examples.

The great tenement building boom in Scottish conurbations in the second half of the nineteenth century saw the sash and case window used almost exclusively. This trend continued into the twentieth century, probably reaching a peak between the wars, with the building of large tracts of social housing. Great housing schemes were built on the periphery of major towns and cities and in smaller pockets elsewhere. The four-in-a-block flatted villa and six-in-a-block tenement were the most common formats. Small-paned sash and case windows were virtually universal in these properties at the time of building.

This author remembers well growing up in a small upper-flatted villa of the late 1930s, in which the 'living-room' window was tripartite. A central section of two small-paned sashes was flanked by two narrower versions, separated by concrete mullions. It is equally remembered that most of these windows were converted to 'picture windows', that is, single-pane sashes, in the 1950s. However, this conversion tended only to be done to the windows facing the street, while the rear was frequently left unchanged for many years.

As building techniques and fashions changed post-war, there was a gradual move away from sash and case to various other types of fenestration. Windows generally became larger, and areas of these were often fixed, ventilation being provided by small side casements or upper-hinged hopper types. The pivot window became common in high-rise developments to facilitate cleaning, but these increasingly required security catches for safety reasons. Despite changing styles and uses, glass can be seen as an integral material to Scotland's built heritage.

CHAPTER 12

PANTILES

Stuart Eydmann

. . . its warm orange or red sits comfortably with local sandstone or limewashed harl . . .

The clay pantile roof is an essential element in the built heritage of Scotland. It is most commonly, but not exclusively, associated with the towns and countryside of the east, where its warm orange or red sits comfortably with local sandstone or limewashed harl and brings welcome colour to flat landscapes and grey coasts.

The pantile was developed in the fifteenth and sixteenth centuries in the Low Countries from older forms by combining the separate lower, flat or concave *tegula* and the upper, convex *imbrex* of the ancient Romans, or the *up* and *over* components of the Mediterranean roof tile combination into an ingenious single element. While there have been many varieties of pantile design, one pattern was the common standard manufactured from the 1600s until the early twentieth century.

Roofing in pantile differs from that of slate or plain flat tiles. Slates are generally laid 'double-lap' fashion, where each is separate from its neighbour in the same course, and each course overlaps with two others to make a roof that is always two slates thick. Pantiles, on the other hand, are laid 'single-lap', overlapping only the very upper part of the course below, the roof being never more than a single tile in thickness. Each pantile also overlaps with its neighbour in the row, the down-bend of the tile on the left fitting over the up-turned edge of that to the right to form a series of shallow channels and raised ridges running down the slope. A new roof is therefore laid in vertical courses from one side to the other, rather than from the eaves up, as with slates.

Pantiles are laid without nails or pegs onto timber battens or 'pantile laths' fixed along the roof parallel to the ridge, each hanging by way of a projecting nib formed on the upper rear of the tile. The tiles are therefore reliant on gravity and the weight of their overlapping neighbours to form a stable, interdependent roof system. In Scotland and England the raised edge is always on the right of the tile, while in the Low Countries tiles were made in both right- and left-handed manners to allow the roof to be laid to suit prevailing wind conditions. The pantile gives a roof that is made from a standard, mass-produced product, is quick to lay, is economic in terms of tiles per square metre, is lighter than one of slate, and is more durable and fireproof than thatch. It also has its own attractive aesthetic qualities.

Technically, pantiles work best on simple, rectangular roofs where a robust and stable system can be achieved. Raised stone skews with a fillet of mortar between the tiles and the masonry give tightness and security, although tiling on the eaves is not unknown, despite this being a more vulnerable and untidy detail. Traditionally, modestly proportioned triangular or three-quarter round-sectioned tiles protected ridges and hips. Valley gutters, where two pantiled roofs meet, were traditionally lined with lead, and both valley and hipped sections required that pantiles be cut to fit. Pantiles were rarely used on pitched dormers, where slates were a more versatile and neater option on such small slopes. One form of dormer that has historically employed the pantile is the 'catslide', a single-pitch dormer set at a lesser angle than the main

A mixture of sources of pantiles, giving
a variegated effect and showing local
variations in the material

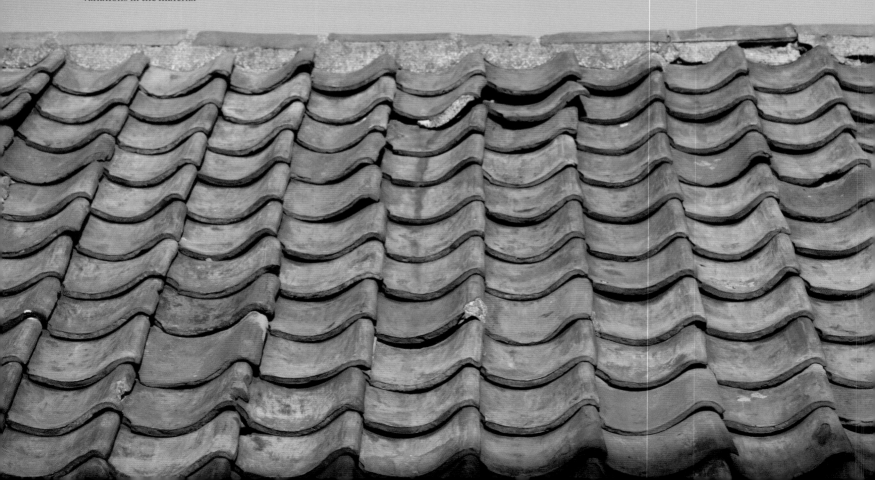

roof; very pleasing dormers of this type can be seen at The Palace, Culross, Fife. Plain and simple forms have visual benefits too. Well-built, square roofs produce neat, straight vertical lines in the pantiles, and setting against the skews with a minimal mortar fillet provides effective visual containment.

Mixing pantiles from different second-hand sources results in a variegated effect, as can be seen on a number of the buildings at Culross. While this might create a picturesque appearance, using hand-crafted products produces technical difficulties due to subtle variations in size and profile. Mixing old and new is also problematic, as contemporary tiles have a different shape from those of the past.

Where a highly windproof and weatherproof roof is required, timber boards or sarking is used beneath the tiles, but in the past this was often eschewed by owners on account of cost or in favour of the benefits of ventilation. Tiles laid without sarking were sometimes daubed ('torched' or 'parged') with lime mortar from underneath to aid stability and improve their seal, and examples have been found of tiles laid directly onto a bed of reeds or mortar on sarking or lath to provide a degree of insulation. Some advocate the careful preservation of such details where found, despite the fact that this can lead to retention of moisture and dampness in the timbers. More recently, an insulating felt underlay has been fixed across the rafters or laid directly onto the sarking to provide a waterproof layer beneath the tiles.

Traditionally, the lower courses were laid to overhang the eaves without gutters, and this produced an attractive wavy shadow line at the wallhead. On some roofs, the lower part of the pitch was flattened or 'bellcast' near the eaves to help shed water. The lower courses were often bedded on lime mortar and, occasionally, a final row of pantiles was set under the second with a deep overlap to protect the wallhead. It is common to find an 'easing course' or 'skirt' of slates or plain tiles at the eaves where, due to the fixed size of the pantile, a full tile's length cannot be accommodated and better protection from rain and snow is required. This may be an earlier thatching practice. Sometimes the slates cover as much as half of the roof, as at the Great House of Pittenweem Priory, or formerly at Fordell's Lodging, Inverkeithing, and this is probably an echo of the fact that the upper section was previously thatched but pantiled at a later date. The easing course is now regarded as standard practice, as can be seen on restored properties at 24 College Street, St Andrews or 4/5 East Shore, St Monans, Fife, and it has become something of an architectural fashion emulated self-consciously by restorers and designers of new buildings seeking 'authentic' detailing. Where roof timbers are out of square, or where the pantiles are loose or do not fit well together, for whatever reason, there is often a reliance on mortar applied to the open areas in an effort to bring stability and a weatherproof seal. This practice has a long pedigree, as can be seen in Dutch and Scottish paintings.

Until the introduction of manufacture by extrusion in the late nineteenth century, pantiles were made by hand. Prepared light clay was cut to size and thickness before being bent over a tile-shaped wooden form or 'horse'. A soft down-curving lipped edge was formed on one side to provide the covering link with the adjacent tile on the roof, the top corner of this edge was rounded to form a shoulder, and the diagonally opposite corner was mitred off to allow the tiles to sit comfortably together. Tiles were left to dry before being fired unglazed. The result was a tile with soft, rounded edges, of obvious hand-crafted character, and offering considerable variety within the standard product. In the words of artist and building restorer Bill Millan:

> All in all the pantile is a fairly rough piece of handiwork. It hides none of its traumas in the making, fashioned out of Mother Earth, lifted and laid with, inarticulate hands well enough versed in its simple construction to repetitively slough with a smoothing thumb here, a hand there and with a swiftness that comes of long experience. Yet for all its vulgar simplicity, pawk-marks and dimensional idiosyncrasies it works.
>
> B. Millan, 'The pantile experience', 2003

Character and interest also came from the local clays, which fired to different colours. Regarding East Lothian, for example, Colin McWilliam writes in *Buildings of Scotland: Lothian except Edinburgh* of the typical 'orange-red over orange-brown walls' and Naismith notes how 'the orange and terracotta pantiles of East Saltoun give way to the redder colours as one moves further east' and 'the gentle toned green tiles of Stenton' (*Buildings of the Scottish Countryside*, 1985). Grey pantiles are found in parts of the Lothians and East Fife, while the

OPPOSITE: An easing course or skirt of slate forming part of a pantiled roof: East Neuk of Fife

tiles of some other areas display the yellow buff of fire clay. The patina is also enhanced by impurities, such as small pieces of lime or sand missed by the milling process which 'blow', leaving a mildly pitted surface, and by the fact that fired clay attracts lichen and moss growth over time, particularly on shallow and north-facing slopes, or where the material has been poorly or unevenly fired.

Despite various suggestions of origins in the Flemish, French and English languages, the word 'pantile' is clearly borrowed from the Dutch *pan*, meaning tile (also *dakpan*: roof tile, *oude holle dakpan*: old hollow roof tile or pantile). This is to be expected, given the geographical origin of the product, although the adoption of the term into English would have been aided by the tile's shallow dish or 'pan' profile.

The patent granted in England in 1636 for the 'maeking of *Pantyles* or *Flanders tyles*, by the way which hitherto hath not beene done by any in this Kingdome' would suggest that the pantile was already known in the British Isles by that time. In Scotland, a licence was sought in 1611 for the manufacture of 'good and sufficient tyill for building and sclaitting of houses', and acts of the Scottish parliament in 1621 and 1681 required that buildings in the principal towns be covered in slate, lead, stone, shingles or tile rather than fire-vulnerable thatch. Whether or not pantile was intended, the product was certainly known in Scotland by 1669, when the Master of Works Accounts refer to pan tayll being used on coach houses. Customs records and correspondence from the last quarter of the seventeenth century show that quantities of *pann tyle*, *pantyll* and *panteijll* were being landed at the Forth ports of Leith and Blackness, principally from Rotterdam.

According to popular tradition in both the Netherlands and in Scotland, the first tiles crossed the North Sea as ballast in the returning merchants' ships that had exported coal, salt and wool to the European ports. However, there is no evidence of ships arriving in ballast, with pantiles used as a convenient and low-value alternative to stones or sand. Furthermore, archaeological evidence from shipwrecked Dutch trading ships has shown pantiles and bricks carefully stowed as part of a balanced cargo, along with other commercial goods.

The late seventeenth century saw the foundation of pantile production in the British Isles. The first

large-scale commercial operation is thought to have been that established by Daniel Defoe at Tilbury in the 1690s for the production of tiles which, as noted by Walter Wilson in his memoirs of Daniel Defoe, were 'hitherto a Dutch manufacture, and brought in large quantities into England'. The earliest reference to manufacture in Scotland is the consent granted in 1709 for the digging of clay to make brick and tiles above Blackness, near Linlithgow, a port already noted for pantile importation and the site of a successful nineteenth-century pantile

manufactory. A few years later, in 1714, William Adam and William Robertson established their works at Linktown, Abbotshall, Kirkcaldy. Their contract to extract local clay for production included payments to the landowner in both money and in pantiles, and it is recorded that they supplied material for the roofing of the offices of Aberdour Manse in 1722, the same year that parliament set minimum dimensions of the pantile (13½ × 9½ × ½ inches) for England. A brick and tile works was established at Prestonpans, East Lothian and by 1716 it

Pantiles create a distinctive
regional character to many
east-coast towns: Anstruther

was supplying material for local industrial buildings while, by 1732, pantiles were being landed at Leith from Stockton-on-Tees (possibly the redistribution of Dutch tiles imported through the port of Yarm) that were claimed to be 'better than any made in this country'. In 1743 the tile works at Leith was offering products which were claimed to be 'as good as any ever come from Holland' and that of Aberdeen, tiles which were 'every way as good as the Dutch'.

Precisely, how and when the pantile came to be adopted as a common roofing material on Scottish dwellings is not known. The burghs of the east coast enjoyed a construction boom during the seventeenth century which saw the building of many new houses and the improvement of others. These included elegant residences and lodgings for the nobility, lairds, sea captains and merchants, as well as many more modest houses and cottages, with their distinctive rubble stonework, lime harl, crowstepped gables and open forestairs. While thatch was their most common roofing material, with stone or slate on the high-status and expensive houses, it might be assumed that the first pantiles found favour with those who had first-hand knowledge of their use in Europe and who sought to emulate their continental counterparts. Subsequently, as tiles became more plentiful and less expensive, they might also have been used on the lower-status houses. In terms of the larger towns, John Slezer's engravings indicate, to some historians, that many buildings in Edinburgh were pantiled by 1690, although it would require further research to confirm this.

There is evidence of the early use of pantiles on non-residential property such as outbuildings, coach houses, workshops, smiddies, tanneries, salt works and even brick and tile works, where they could provide an economic, fireproof, dry and well-ventilated roof. One major industrial project, the New Mills Manufactory near Haddington, East Lothian,

was designed in 1695 to be roofed in 24,000 pantiles, which would be supplied by a merchant trading with Holland. It is recorded that they were used on the outbuildings of a Fife manse in 1708 and at Airth, near Falkirk, where it was noted that pantile was commonly used to cover 'offices'.

The pantile found use on the houses of the workers of the emerging rural landscape, as in the new villages of Charlestown, Fife and Ednam in the Borders. However, it has been suggested by Fenton and Walker in their *Rural Architecture of Scotland* (1981) that the covering of rural houses with pantiles caused 'considerable controversy' and was relatively rare prior to the 1850s, unless used as an inexpensive substitute for thatch.

This gradual if uneven replacement of thatch with pantile or slate is well illustrated in David Wilkie's 1840 painting *Pitlessie Fair*, which has as its central background a cottage in the Fife village which has been recently pantiled, but with turf or thatch still showing at the ridge. The nearby villages of Falkland and Auchtermuchty retained some thatched buildings well into the twentieth century, while others in the vicinity were re-roofed in pantiles from the works at Dunshalt.

Having found its place in Scotland, pantile manufacture was greatly stimulated in the early nineteenth century as agricultural improvements brought the new architecture of carefully designed steadings and associated buildings, which have come to typify much of the Lowland landscape. The tiles were recognised as an economic means of covering the large new structures and brought advantages of ventilation to animal housing, while slates were used on the residential buildings and those where more weatherproof conditions were required. Effective drainage of land was a major aim of the improvement programme, and both estates and private entrepreneurs established new works where

tile drains and roof tiles could be manufactured together. Horse and later steam power was used in the processes, and works were usually located close to good sources of both clay and coal. Transportation of fuel, raw materials and completed products was also facilitated by water, as at the works at Charlestown, Fife, established around 1780, which offered coal and lime as well as tiles; or the Blackpots brick and tile works in Banffshire, established around 1785. Both of these had their own harbours. There were three works adjacent to the Forth and Clyde Canal at Falkirk, the largest owned by the Earl of Zetland, while on the Monklands Canal works, erected in 1785, manufactured both pantiles and 'slate tiles' in imitation of true slates. Production was encouraged by the removal in the late 1830s of a Brick and Tile Tax, imposed in 1784, and the establishment of brick and tile works built to a peak over the first decades of the nineteenth century, so that by 1869 there were 122 operating in Scotland.

Berwickshire was served by tile makers at Berwick and Coldstream, while East Lothian had works at Musselburgh, Portobello, Prestonpans, Longniddry, Cousland and elsewhere. The brick and tile works at Blackness, writes Samuel Lewis in *A Topographical Dictionary of Scotland*, had the benefit of 'a valuable field of clay, twelve feet in depth' and by 1834 'its produce amounted to 150,000 bricks, 200,000 roofing, and the same number of draining tiles' with further increases over the next decade. No doubt many of the handsome farm buildings of the landscape immediately to its south, including those on the Hopetoun Estate, are roofed in tiles from this works.

The fact that the agricultural improvements were initially focused on the east coast helped consolidate the market there, sustaining local skills in both making and laying and reinforcing the pantile's position in the local architectural vocabulary. In the countryside of the west, where access to slate was easier and rainfall heavier, the pantile never took hold to the same extent. John Naismith noted in *General View of the Agriculture of Clydesdale* that although many pantiles were being made and used in the county, they had a bad reputation on account of the poor quality of the first tiles made, while another commentator writing of Ayrshire noted that tile roofs did not last in that county. Nevertheless, there were substantial roof tile works at industrial Wishaw and Coltness, Lanarkshire.

As the nineteenth century progressed, builders and property owners in the rapidly expanding towns and cities came to recognise the opportunities offered by the pantile. Drawings, prints and early photographs show that many older thatched or slated buildings, some of which were in an advanced state of decay, were given prolonged life under pantiled roofs, and the pantile was commonly employed on workshops and commercial properties. In Edinburgh, for instance, it can be seen that large numbers of old houses in the closes and wynds off the High Street, Canongate and Cowgate had been patched in tile, and that cottages, stables and industrial premises on the fringe of the city had many pantiled buildings. Also, photographs of once high-status structures such as Cardinal Beaton's House, Edinburgh; Lamb's House, Leith; Mar's Castle, Aberdeen; and many other 'palaces' recorded in the drawings of MacGibbon and Ross had also been pantiled by the second half of the nineteenth century. Even Glasgow had pantiled buildings. There was a tile work there by 1778, and a 'pantyle-maker' was recorded at a city centre address in the 1780s. Of the Burgh of Calton we read in Lewis' *Topographical Dictionary* that 'the houses are neatly built of brick, and roofed with tiles, for the manufacture of which clay of good quality abounds in the immediate vicinity'. This was the same place where, in 1819, a bylaw was introduced to prohibit the roofing of houses with thatch, tar or paper. Historical images

OPPOSITE: Stonework, lime harl and crow-stepped gables combine with pantiles to create a distinct building type

show that pantiles were less common on tenements or city buildings over four storeys, presumably on account of vulnerability to wind and the danger of falling materials.

Inland towns, including Bathgate, Dunfermline, Dalkeith and Linlithgow, as well as coastal settlements such as Bo'ness, Musselburgh, Dunbar and Stonehaven, all had a high proportion of pantiled roofs in the nineteenth century, although the balance has now tipped in favour of slate, as it was precisely these buildings that were the targets of slum clearance and comprehensive redevelopment schemes. By the end of the nineteenth century, the pantile was regarded as suitable only for temporary and inferior buildings, such as sheds and outhouses.

In parallel with these losses, the search for a national architecture based on historic precedents, combined with the emerging conservation movement, recognised and privileged the pantile as an authentic element of the built heritage to be both conserved or reintroduced. For example, pantiles were introduced at 443–9 Sailor's Walk, Kirkcaldy sometime in the 1920s or 30s, prior to the later 1954–9 reconstruction by the National Trust for Scotland.

OPPOSITE: Pantile was used to roof many dwellings, in this case that of a fisherman

BELOW: A rare example of a circular roof covered in pantile

152

154

A traditional pantile roof with modern replacement in the background

Hurd and Neil's 1937–9 conservation work at Lamb's House, Leith saw asbestos sheeting replaced with pantiles, and tiles were introduced in the conversion of the seventeenth-century 189–191 Canongate, Edinburgh. The book *Rossend Castle* records that, in the high-profile 1970s' restoration of the roofless castle at Burntisland, Fife: 'pantiles, rather than slates, were perhaps surprisingly selected for the major covering but the castle is on the coast where pantiles are prevalent. It seemed that they were an appropriate roofing material and their use perhaps also provided a little side step from the conventional approach.' John Knox's House in Edinburgh was pantiled in 1981, despite the lack of any evidence that it ever had such a roof. More recently, Queensberry House, Edinburgh, was given a pantile roof for the first time as part of

its incorporation into the complex of the Scottish parliament, although only after some debate.

Modern pantile work, undertaken with due care and attention to traditional practice, can be seen in the restorations of the National Trust for Scotland's Little Houses Scheme. It is also carried out in the projects of some local authorities and building preservation trusts working with grants, and within the listed building and conservation area regimes. The pantile has been used successfully on modern infill and replacement buildings in historic areas of the east-coast towns and cities, as in the work of Basil Spence and Robert Hurd in the 1950s and 1960s, and on new housing where a regional identity drawing on the traditional vocabulary of simple forms and detailing was sought.

Native pantile manufacture dropped off during the early twentieth century, accelerated by a decline in demand for clay field drains. Despite some attempts to reintroduce manufacture there are now no Scottish tile works, and most products now come from a single supplier in Humberside. Today tiles are made by extrusion and, while they perform well, they lack many of the subtleties of the old models, including their soft edges and handmade character. A building roofed in modern tile also has a different appearance as, in the new products, there is a greater emphasis on the convex roll as opposed to the shallow pan section in older tiles. There is also a modern preference for heavier 'high-performance' tiles with interlocking parts and these, too, have a blander, less interesting effect. Cheaper concrete pantiles are also popular.

It is very difficult to read the geographical origin and date of tiles but, assuming that the expected life of a pantile is around 150 years, it can be assumed that most of the tiles imported into or manufactured in Scotland in the seventeenth and eighteenth centuries no longer survive. It is possible, however, that some very early examples exist in roofs made up from material salvaged from older roofs, as at Culross, and there may still be some unimproved buildings that retain early tiles preserved by neglect. Furthermore, those pantiles, which were manufactured during the peak of production and use in the nineteenth century, must now be coming to the end of their useful lives. While many remain on farm buildings, a large number of these structures are redundant or at risk, and in towns most pantiled roofs have now been renewed with modern products. Those original tiles that survive therefore constitute a rapidly diminishing resource which, like Scottish slate, is worthy of recognition, salvage, conservation and sensitive reuse.

However, the pantile is enjoying a new lease of life in current construction. This is well illustrated in two projects: Robert Steedman's Muir of Blebo, Fife, uses sandstone and pantile in a modern house that has the character of a much older building; and at Robert Potter and Partners' crematorium at Roucan Loch, Dumfries, the prominent pantile roof suggests tradition and rootedness, while bringing warmth and texture to a sensitive public building and landscape. As these show, the humble pantile still has the potential to contribute to the Scotland's future built heritage as well as to that of its past.

155

The rich colour and texture of a pantiled roof

CHAPTER 13

THATCH

Tim Holden

. . . thatching techniques tend to follow strong vernacular traditions.

Introduction

The thatched house is an endangered species in Scotland, but roof coverings utilising available plant resources have undoubtedly been used since the first settlers arrived in Scotland. In much of rural Scotland this was still the case well into the nineteenth century, until improved systems of transport, in particular the railways, gave better access to cheaper, more durable materials. Sheet materials such as corrugated iron, and pantiles, slates and wooden shingles, were suddenly a viable alternative and, following the early lead of the towns, rural buildings of thatch, turf and hewn timber began to disappear.

The demise of the thatched house has dramatically changed the character of Scottish settlements. In more remote areas, thatching as a roofing technique tended to survive longer because of a lack of affordable alternatives, but even here the depopulation of the countryside and the general improvement of housing standards have reduced its viability. Alongside the perception that thatched roofs signal rural poverty, the gradual loss of vernacular skills and the increasing difficulty of sourcing materials has meant that these skills have slipped away from us largely unnoticed.

Thatched buildings that were once dwellings were commonly demoted to animal houses, and the levels of investment required for their upkeep was reduced as a result. Unless a thatched roof is actively maintained it quickly falls into disrepair. The thatch absorbs water, the timbers rot and collapse quickly follows. This need for constant care, together with the expense of periodic rethatching, helps to explain why there are so few surviving thatched houses in Scotland. Of course, there are many standing buildings that show signs of having been thatched in the past, but these add little to our current understanding. By far the best materials for study come from the remains of thatch that survive preserved beneath sheet materials such as corrugated iron or asbestos. This, too, is a diminishing resource, but recent work carried out in advance of demolition or natural collapse has helped to illuminate quite how diverse these roofs can be in terms of both materials and the techniques used. This type of evidence perfectly complements the documentary, pictorial and photographic evidence upon which we traditionally rely so heavily.

A traditional Western Isles thatch, showing rounded roof profile

Regional variation

Because of the relative remoteness and rural character of much of Scotland, thatching techniques tend to follow strong vernacular traditions. Most rural buildings would have been roofed and maintained using local materials by the owners, their tenants or members of the local community.

Regional styles depended very strongly on the environment, having been developed over many years as the most suitable to cope with local conditions. But other variables also played an important part in determining which materials were used initially and how the buildings were maintained through the years. Availability of manpower was an important variable that changed periodically as the younger men were diverted to war service, the merchant navy or by the lure of work in the bigger towns. As the economic circumstances of the community or the occupants changed, so did the ability to acquire the best materials and provide the annual attention these roofs needed. At the same time, especially since the Second World War, rural areas also saw a radical change in agricultural practice through the introduction of regional specialisation, chemical fertilisers, improved cereal varieties and changes in grazing patterns. All of these had an adverse impact on the availability and quality of thatching materials to many households.

The authors of *Thatches and Thatching Techniques* (1996) suggest that Scotland can broadly be divided into seven different areas with respect to the types of thatching techniques used. On the exposed Western Isles, for example, thatched houses exhibit low rounded profiles, typically using cereal straw or marram grass fixed in place with intricate roping techniques, fishing nets or, in more recent times, by chicken wire. In these coastal areas a longer growing season puts less pressure on cereal straw for fodder, and this can be used instead for thatching farm buildings, as well as corn and hay stacks. The lack of readily available timber means that alternatives such as driftwood and straw rope are commonly used to support and secure the low profile needed to withstand the high winds and driving rain.

By contrast, in the Highlands, it is important that a roof can quickly shed rainwater and snow, and the pitch of the roofs tends to be steeper. Here, in the short growing season, oat and barley

A thatch of broom: although harder to work with, such natural materials were durable and allowed agricultural material to be used as animal fodder

straw is a valuable fodder resource, and the more durable heather, broom, and bracken are used as an alternative. Although more difficult to handle, these are readily available and are also less attractive to browsing livestock and deer.

Knowing more about the context in which a thatched building stands can explain why certain techniques were used rather than others. Conversely, a detailed examination of existing thatched roofs can also inform us about such things as the status of the occupier and their socio-economic environment.

OPPOSITE: Driftwood used as a basal layer for a Western Isles thatch

BELOW: Heather rope used as sub-stratum, Orkney

Materials and methods

A thatched roof can have a considerable weight and needs to be strongly supported. Timber is used in the vast majority of cases, although other materials, such as whale bones, are also incorporated where available. The structural members usually support smaller-diameter wood (cabers) or sawn offcuts (backs) running down the pitch of the roof. Where wood is scarce, other methods have to be devised. One solution used on Orkney is the so-called 'needle thatch'. Here, close-set lengths of straw rope (simmens) are tightly, and repeatedly, stretched between the ridge and the eaves to support the straw thatch above.

In some situations thatch is used in conjunction with other construction materials, such as flagstones or pantiles. In these cases, the main function of the thatch is to direct the water away from the joints and to prevent the upward movement of water between seams in high winds.

Most thatches comprise three main elements: a basal or sub-stratum layer; the main body of the thatch; and fixings.

Basal or sub-stratum layers

The basal layers often differ in character from the body of the thatch and perform various functions depending on the type of roof. They prevent the overlying materials from falling into the interior and provide a good seal with the gable or wall heads. They also create a level bed onto which the main thatching materials can be fixed. Basal layers include heaped vegetable material used as stuffing/levelling layers, and supportive structures such as timber, brushwood, turf, wattle and straw rope.

Main body of the thatch

A wide range of materials were used to thatch Scottish buildings. Turf was sometimes used on

RIGHT: The Highland cottage, Aberfoyle, symbolic of a lost vernacular tradition

OPPOSITE: Random thatch

animal houses or temporary shelters such as shielings, but the main thatching material was usually of plant origin. Both wild and cultivated species were used, sometimes together on the same roof. Cereal straw, including oat, barley, wheat and rye, were common materials depending on availability. Rye, although not generally considered as a staple cereal, may have been cultivated specifically for its long thatching straw. Other cultivated species likely to have been used include flax stems, bean stalks and potato shaws.

Wild species were also widely used where cereals were scarce or required for other purposes. Bracken, heather, marram grass and broom thatches are relatively common throughout Scotland. The following are all known to have been used:

Native grasses (e.g. Couch grass - *Agropyron repens*)
Reed (*Phragmites communis*)
Rush (*Juncus* spp.)
Iris (*Iris pseudacorus*)
Sedges (*Carex* spp. – particularly *C. pendula*)
Bracken (*Pteridium aquilinum*)
Dock (*Rumex* spp.)
Heather (*Calluna vulgaris*)
Juniper (*Juniperus communis*)
Broom (*Sarothamnus scoparius*)
Seaweed (Various species)
Eel-grass (*Zostera* spp.)

B. Walker et al., *Thatches and Thatching Techniques*, 1996

These plants would have been collected locally, and availability would have been highly dependent on the local ecology and land management practices. Some may have been managed specifically with thatching in mind.

Jim Souness, in an article on heather thatching in Scotland, describes the best type of heather for thatching as being long (up to 4 feet), straight, unbranched and often to be found growing on steep north-facing slopes. Areas like this that provided better quality materials would undoubtedly have been given a degree of protection from grazing or burning by the owners. Some traditional land-management techniques would also have inadvertently produced better quality materials. Regular cutting of species, such as marram, for animal feed would, for example, have produced thicker stands that were better suited to thatching than their unmanaged counterparts.

The depopulation of the countryside and changes in agricultural practice have contributed in no small way to the decline in thatching as a rural craft. Nitrogen-rich fertilizers and more mechanised harvesting and threshing techniques generally result in poor-quality cereal straw for thatching, and the burning of heather for grouse shooting are some of the practices that have had knock-on effects on the quality and quantity of available thatching materials.

There are commonly two ways that thatching materials are applied to a roof. They are either placed with no consideration for the direction of the stems, or an attempt is made to ensure that the materials are aligned, with the stems broadly running parallel to each other down the pitch. For obvious reasons, the two techniques are described as either *random* or *parallel* respectively.

A random thatch requires very little preparation, but the water-shedding properties are poorer, and water undoubtedly penetrates

162

further into the roof as it trickles towards the eaves. This type of construction is typical of the Lewis Blackhouse, and in both the Western and Northern Isles it is typically associated with the intricate roping techniques and annual maintenance that is needed to counter high wind speeds and driving rain. It would seem likely that the presence of a permanent fire in the hearth and the passage of warm air through the thatch are essential elements in keeping the roof dry and in good condition.

In parallel thatches it is much easier to see how the water is directed down the aligned stems, dripping from one to another and arriving at the eaves before it has penetrated too far into the roof. Much more care is taken in the preparation of the materials: leaves and weeds that might encourage decay are removed, materials might be raked before fixing, and more care is taken over the final dressing. Once in place, less day-to-day maintenance is required, but more time needs to be expended in construction, which probably requires a greater reliance on specialist craftsmen.

Fixings

There is a wide variety of methods for fixing the thatch onto the roof, many of which are discussed in detail in *Thatches and Thatching Techniques*. In some kinds of thatch, the fixings are concealed within the main body, but in others they are visible on the surface.

Concealed methods include several stitching and pegging techniques and the use of turf and prepared clay to hold the thatching materials down. These may be augmented in exposed areas by external fixings.

Surface techniques include the use of turf, clay, wooden pegs, twine/rope, netting, wire, timber or metal stays, and wooden staples. Special attention has to be paid to the most exposed parts of the roof, the ridge, the skews and the eaves. Here additional materials such as broom, bracken and turf are used in

conjunction with weighted ropes and netting. Some methods of thatching involved the application of clays, dung or mud.

OPPOSITE: Parallel thatch

BELOW: Weighted ropes and fishing net used as a fixing for thatch

165

Lifespan and repair

The durability of different classes of thatch varies considerably according to materials and the environment. A good heather or cereal thatch in a sheltered location might, for example, last for more than twenty years. But in many parts of the Western and Northern Isles, an annual dressing of fresh cereal straw would be needed on a fully maintained roof. At Keils on Jura, the owner of a building thatched with rushes recalled that he and his father needed to apply a surface coat to alternate pitches of the roof in consecutive years.

The repeated resurfacing of the roofs over many years inevitably resulted in a build-up of material that could eventually exceed 1 metre in depth. On old photographs, this process can be seen, as both the profile and height of the ridge changes through time. A recently studied thatch from Gimps on Orkney appeared to show evidence for up fifty rethatchings, each of which had been compressed into a narrow band.

Born in the thatched cottage, this former 'owner–occupier' was well versed in its upkeep: Western Isles

By contrast, a thatch from Sidinish on North Uist clearly showed that there was a basal layer of turf, overlain by a lower stratum of marram, rush and cereal straw, with several layers of heather each up to 20cm in depth. Each layer was defined by an eroded surface below traces of the heather rope fixings. On roofs such as this, periodic overhauls were required when much of the old thatch was removed and the remainder resurfaced.

OPPOSITE: Thatch was often to be found used in conjunction with other vernacular techniques, in this case cruck-framed construction

The archaeology of thatch

The maintenance and renovation of the few surviving Scottish thatched buildings represents a dilemma for all concerned. Any significant renovation, whether of an exposed thatched roof or of a part-preserved structure beneath corrugated iron, will require an element of downtaking. Although this is essential for the long-term survival of the building, the downtakings also potentially destroy valuable evidence about the methods and materials used. Several recent projects funded by Historic Scotland have required the detailed archaeological recording and analysis of threatened buildings in different parts of the country. Of the nine surveyed to date, no two buildings used the same techniques, an indication of the diversity of vernacular tradition and also a warning about how much information could be lost if these types of building are allowed to disappear unchecked.

In addition to evidence relating to the roof itself, the materials within the thatch can provide evidence relating to social history, agricultural practice and the environment that is often difficult to obtain from other sources. The shape of surviving roof turves, for example, reveal the way they were cut and the tools used. They also provide very detailed information regarding the plants that grew on them and the type of ground they were stripped from. Both of these offer direct evidence about land availability, the local environment and the way it was managed. Often whole cereal plants are recovered, and these can be used to provide information on the crop varieties (many long extinct), the local economy and the ecology of the fields in the nineteenth century and even earlier. Evidence for how the crops were harvested (by uprooting or cutting) can be determined, while the weed seeds included with the crop can offer an insight into the conditions in agricultural fields in the years before the routine application of herbicides and chemical fertilizers.

Finally, the data recovered from old thatched roofs also provides detailed information that is invaluable to archaeologists in their interpretation of sediments from long-abandoned settlements. Their potential for providing relevant information in many spheres of academic study should not be underestimated.

Summary

Thatching, in its various forms, has undoubtedly been used as a roofing technique since the very first people settled in Scotland. Whether on temporary shelters or permanent dwellings, a wide variety of plant resources together with turf, clay and the like have been used to protect us from the elements. Many different vernacular traditions have developed over the years in response to the environment, local materials, economic circumstances and available technologies. These traditions helped to provide different regions in Scotland with their own distinctive character that would have been readily recognised by occupants and visitors alike.

Thatched housing has been under threat for many years as other roofing materials that require less maintenance became more readily available. During the post-medieval period towns and villages gradually saw new buildings constructed with pantiles and slates, and many of the older thatched roofs were gradually also replaced. In rural areas the story is similar, although thatching was still commonly practised well into the twentieth century. Here, many buildings were eventually roofed with sheet materials such as corrugated iron or asbestos. Some still retain the remnants of the thatch beneath the current surface and, in the absence of many fully maintained thatched buildings, these now represent an important heritage resource.

CHAPTER 14

SLATE

Neil Grieve

Slating is a traditional building skill, which looks superb and brings real individuality to many of Scotland's buildings.

William Morris, writing in 1890, expressed the opinion that 'There is nothing more important in the aspect of the exterior of a building than the covering of its roof '. Among these good roof coverings, he listed grey and dark grey slates as were used in parts of Scotland, which he described as being 'small and thick'. Morris' description of the material is well meant, but for anyone who has no knowledge of, or interest in, Scottish slate it could be seen as slightly disparaging. It might even contain a negative implication about the manner in which the slate is laid, whereas in reality, Scottish slating has a long provenance, possesses real subtlety, and has had a marked effect on the country's architecture. Good Scottish slating has every right to be considered a craft rather than a trade.

In Scotland, by the fifteenth century, craftsmen were being described as slaters. The Exchequer Rolls of 1427–8, for example, mention a payment made to William de Law and his partners, who are described as 'Sklaters,' for stone slates used at Linlithgow Palace. By the late 1500s the 'master slater' had arrived, indicating that a reasonable volume of work was being undertaken. In Dundee, where the slating trade appears to have had a particularly early eminence according to their minute book, the slaters were incorporated into the three united trades by 1684, suggesting that the trade was well organised and reputable. By the mid-1700s, the Scottish building tradition had evolved to the extent that buildings, particularly in the burghs, were expected to be stable, permanent and fire-safe. Bylaws insisted that new houses had to be given a slate or tiled

172

roof, and lined chimneys were introduced. Clearly, by this time slate was reasonably easy to obtain. In fact, from the later 1600s 'scailzie', or 'skailie', which was initially used to describe any flat plate of stone, tile or timber suitable for slating, begins to be described as 'black', indicating the use of true slate as opposed to sandstone or some other material. In combination with stone, harl, limewash, slabs, cobbles and setts, slate replaced the old timber tradition in building and contributed to a new national aesthetic.

The word 'slate' was, and to an extent still is, something of a generic term applied to any stone used for roofing. Initially, the preferred material was sandstone, although schists and shale slates – and, in fact, any stone which split easily into thin sheets along its bedding planes – was used. Fissible (easily split) sandstone is available across Scotland and was originally extracted in some quantity, to an extent that is greater than might previously have been suspected. Sandstone slates were usually referred to as 'brown' or 'grey', which distinguished them from true slate. The form of the few surviving medieval or late-medieval roof structures, which tend to be slightly uneven with open battens (Claypotts Castle, near Dundee, for example), suggest they were constructed to take the weight of heavy stone slates. Drummond Castle near Crieff, which is some distance from any known source of stone slate, still retains them on its keep. Caithness in the north of Scotland and Carmylie in Angus became the main areas of production, and while both achieved industrial status, by the 1820s true slate had come to predominate and the sandstone quarries concentrated on producing flagstone, for which there was commercial demand.

At one time, Scotland had at least eighty slate quarries, the majority of which were related to crofting work and only supplied local needs. Less than thirty of them were run as commercial enterprises, and by the early 1800s the Scottish slate industry centred

OPPOSITE: An expanse of Scottish slating showing diminishing courses and random widths characteristic of the Scottish tradition

BELOW: Scottish slate used to roof Stirling Castle

on four localities which produced material markedly different in quality and appearance. They were:

1. The islands of Easdale, Seil, Luing and Belnahua. Because it was so easy to ship the material out by sea, the island of Easdale was probably the earliest source to be worked with any consistency in Scotland. John White, manager of the quarry from 1840 until 1865, wrote in the Mining Journal in 1864 that 'Caisteal-an-Stalcaire or Falconers Castle at Appin was roofed in 1631 in Easdale slate'.

2. Ballacullish, on the shores of Loch Leven at the head of Glencoe. First opened in 1693, this became the biggest slate quarry in Scotland. The material was of excellent quality, relatively easy to work, and was also shipped by sea all over Scotland. By the mid-1800s, Ballacullish was producing some 25 million slates annually and had supplanted Easdale as the main area of production. There is some controversy over who first opened Ballacullish: men from Easdale or the Lake District. Dorothy Wordsworth, in her journal of 1803, reported

OPPOSITE: Slate covering a bell-shaped roof: Stanley, Perthshire

BELOW: The impact of slate quarrying can clearly be seen here on Easdale Island
© Crown Copyright: RCAHMS

OPPOSITE: A range of different dressing styles are employed here to create a decorative effect

meeting men from her native Cumberland at the quarry, and she commented on the large number of ships in the harbour.

3. The Southern Highlands, close to the Highland Boundary Fault. Worked in a series of quarries literally stretching from coast to coast, this might initially have been the most worked area, providing material mainly for local markets; these included the roofs of towns such as Stirling and Perth.

4. The Macduff and Foudland quarries in the north-east of Scotland. While not so well known, slates from this area enjoyed a good local reputation and were used freely by local architects and builders. The Macduff and Foudland quarries had ceased working by about 1870, all of the others by the mid-1950s.

During a peak of production in 1895, Scottish quarries produced 45,000 tons, which was approximately 8 per cent of the national total. By 1914, this output had dwindled to 8,000 tons. The demise can be attributed to a number of factors, among which high cost, lack of investment, shortage of labour, and the dominance of imported slates and concrete tiles were the most crushing. Ironically, a second-hand market had always existed and quickly strengthened, testifying to the merits of the material.

The difference between true slate and the other materials is that the latter will split along bedding planes, while the former splits along different planes known as 'slaty cleavage'. True slate originally existed as fine-grained muds or silts. As sedimentary layers built up, weight caused the beds to compress to form shales, which were then subjected to further intense geological pressures exerted by high temperatures and lateral movements in the earth's crust. This folded the rock and converted most of the clay minerals into white mica and chlorite, which then realigned themselves in one direction (the grain). They overlapped like fish scales, which gave the rock exceptional strength and its 'slaty cleavage', or the ability to split easily.

In addition to micas and chlorites, slate also contains varying amounts of quartz and small quantities of materials such as iron ores, calcite, magnesite and graphite. Because it is a composite material formed under varying conditions, it varies greatly in both its appearance and physical properties. Broadly, the ratio of the platy materials to the hard quartz particles determines how easily the slate will split and whether it will be, for example, thin and smooth or coarse and thick. Colour can vary depending on the iron and carbon content of the slate, and colour banding might be inherited from the original bedding planes. Larger grains of quartz may be squeezed into 'knots' or 'eyes', and may also be responsible for the striation which sometimes features on the surface of the slate. Pyrites crystals (referred to as 'diamonds' by the quarrymen) are prevalent in certain seams. However, so long as slaty cleavage has occurred, the slate that has been formed is capable of providing a good roofing material.

During folding, fractures or joints form in the slate, and the spacing of these joints and the angle at which they cross the cleavage plane determines the size of slates that a quarry can produce. To produce larger slates, a higher wastage ratio will result, whereas if small slates are acceptable, the amount of material recovered will increase. As the slate industry across England and Wales mechanised (powered circular sawing was common in Wales by the 1850s), larger blocks became easier to produce. Around twenty standard sizes, based on the 'Female Nobility' sizes – for example, Duchesses (24 × 12 inches) and Countesses (20 × 10 inches), introduced by Colonel Warburton of Penrhyn – became the norm.

By way of contrast, Scotland, even by the early 1900s, had achieved nothing like the same degree of mechanisation. Quarry working was based on the long-established practice of owners letting a section of rock to a crew of five or six men, who would then work that area for a year while being paid an agreed sum for each 1,000 slates they produced. The crew divided into three trades: quarrymen, responsible for bringing the slate out of the quarry; splitters and dressers, who split the rough blocks into thin sheets before trimming or dressing the slates into the most appropriate length and breadth; and labourers, who moved the slate blocks from the rock face and disposed of the waste. Irregular jointing, lack of machinery, and a system of working that encouraged maximum recovery meant that the Scottish quarries, while capable on occasion of producing large slates, did not really have the option of sorting their production into regular sizes. They continued to deliver random sizes in different thicknesses, usually classed as 'sizables', which averaged 15x8 inches, and 'undersized', which averaged 10x6 inches. Length was always at least twice the width.

Despite one or two attempts to export the material, the output of the Scottish quarries was intended for home consumption. Almost everywhere else in the British Isles moved over to regular-sized or 'tally' slates; and for a time, until the English and Welsh quarries became fully mechanised, Scotland became essentially a dumping ground for their random sizes. Scotland has always been a net importer of slate.

Because slates will more easily fracture parallel to the grain, it is important that the grain does not lie in the direction of the shorter width. The dressers were well aware of this, and one reason for the common practice of shouldering – that is, rounding the top corners of the slate – would be to indicate the way in which it should be laid on the roof. Good slate, little affected by heat, damp or frost, is arguably, in terms of its weathering qualities, the best of the natural building materials. Crucial to its longevity, however, is the manner in which it is laid.

The purpose of slating is simple. It is to lay a network of flat plates, calculating the manner in which they overlap in such a way that water will run, under the force of gravity, from the ridge to the wallhead without penetrating into the roofspace. However, people rightly expect their roofs to last, and because roofs can be large, accounting for a surprising volume of a building, and can also be highly visible, with eye-catching features such as turrets and dormers, to this basic definition should be added some recognition of the ideal of achieving a degree of permanence and an aesthetic quality – both properties which characterise Scottish slating.

The essential difference between Scottish practice and the way in which slating is undertaken elsewhere can be attributed to two interwoven factors. These are the harsh, at times extreme, climate; and the comparatively small, randomly sized material produced by the quarries. While Morris may have stretched a point when claiming that the roof covering is the most important aspect of the appearance of a building, in many cases the roof is in plain view, and these twin factors not only affected the way roofs look, but the way in which they affect the architecture as a whole.

The more extreme the climate and the more exposed the situation, the greater the pitch of the roof. A steeper pitch is less prone to allowing wind-driven rain or snow to penetrate the roof space and will move water quickly down the roof, thereby minimising the potential for water to penetrate between joints and fan out under the slates. The traditional steep pitch means there is headroom in the attic space which, if the plan of the building is deep enough, is frequently inhabited. This tends to give rise to interventions in the form of, for example, dormers, and encourages

OPPOSITE: Stone slates, Orkney

179

the provision of hips and valleys, the angle of which is less than that of the main roof, making these details more vulnerable to the elements. Once nailed in position, a slate will angle up over the slates in the course below, as a consequence of which its surface will be some 5 degrees less than the roof pitch. For all of these reasons, Scottish roofs are usually steeply pitched, with 40 degrees being a common average.

The small slates were ideally suited to steeply pitched roofs. A single nail hole at the head of a slate was sufficient to hold it in place, and while slates fixed in this way do not have the benefit of a long lever arm above the nails to resist wind uplift, as in centre nailing, their weight and small size, which offered a lower degree of wind resistance, kept them in place. Shouldering and single nailing enabled slates to be swung aside to expose the nails of the slates in the courses below. In the event of a slate coming loose, therefore, it could easily be re-nailed. This useful property is improved by the practice of laying the slates in diminishing courses: that is, starting with the larger slates at the eaves and gradually reducing the size of the slates, course by course, until the smallest slates are used at the ridge. This process enhances the visual impression of steepness, particularly when, as is common with many Scottish buildings, there is no cornice or other detail at the wallhead. Also, the variety of sizes made covering the conical and pyramidal form of turret roofs much easier, thereby encouraging their use. All of this adds significantly to the effect of the skyline signature of many of Scotland's buildings and much of its townscape.

The use of slate was for a long time inhibited by the cost of producing sarking. Battens were relatively easy to produce and were suitable for tiles or large relatively standard-sized slates but were not so good for small decreasing courses. Offcuts from pit-sawn timbers were used, but by the late 1700s sawn boards had become more readily available, making the process of laying a large number of randomly sized slates very much easier, there being no need to calculate batten widths. It also had the advantages of greatly increasing structural stability, providing draughtproofing and insulation, and presenting some resistance to water penetration.

The slater, once his slates were delivered to the site, would sort, size and hole them. He would probably be selective about his choice of slate, and might originally have carried out his own tests, which usually involved a period of immersion in water. Slates would arrive shouldered but without a nail hole, which the slater would make centrally at the top of the slate, about 1 inch from the head. Nail holes are punched through from the bed and should cause a small amount of spalling on the back, which acts as a counter sink for the nail head. Slaters preferred to hole slates themselves, as it gave them an idea of the quality of the material they were working with.

When sizing, the slater would need to work out roughly how many slates would be required per course and then ensure that he had enough slates of a particular size. He would want to be certain that the courses diminished evenly: an obvious jump in sizes between courses was a sign of poor workmanship, and too many courses using the same size of slate was also considered bad practice. There is little doubt that the aesthetic quality was important to a good slater. Sizing is carried out using a pin rule, so called because it has a nail or 'pin' at one end and is numbered, usually at half-inch intervals by lines drawn across its length. The slate is measured by sitting the tail on the pin and referring the nail hole to the nearest numbered line. While the size will invariably be taken from the nail hole to the leading edge or tail, the system of measurement does vary across the country. The most common system has 11 inches as its benchmark. An 11-inch slate is a size eleven; thereafter size increases or decreases by half an inch at a time, so a 10½-inch slate is a ten; a 10-inch slate is a nine; while, increasing in size, an 11½-inch slate is a twelve; a 12-inch slate a thirteen, and so on.

Slates would also be graded according to weight, usually into the three categories of thick, middle and thin. Slaters could make use of the variety in weight in a number of ways. Some would simply mix them, believing that the difference in thickness would add to the attractive natural appearance of the roof. Others might use the thicker slates at the skews or verges and then move into the centre of the roof, using the middle thickness next, and finally the thinnest. This very slight dishing effect, not really discernable to the eye, would encourage water away from the vulnerable skews or wallheads, a process known as 'tifting', the 'tift' being the gap between slates and the wallhead on which they sat. Some slaters would retain the thick slates for use where there might be a risk of mechanical damage around dormers or in the vicinity of chimneys. Ideally, wider slates equating roughly to one-and-a-half times the average size should be set aside for use at the ends of courses, to avoid having to use half-width slates.

The head lap or cover is the most important calculation the slater makes. It is usually 2–3 inches, decreasing to under 2 inches at the top of the roof to allow the smaller slates to lie properly. The slater will calculate the margin or guage of each successive course by subtracting the cover from the length of the slate and dividing by two. The varied widths of slate meant that it was impossible to maintain an even side lap. Because of the uneven surface of many Scots slates, the tendency for water to fan out under the slates under capillary action is not so great, but the side lap is still important and, as a rule of thumb, should never fall below 2 inches. On particularly wide slates, a narrow in-band would be placed in the centre to take up the additional width.

Slate dressing at the quarry
© National Museums of Scotland

It was usual to cheek nail every third course to help keep slates in place, and T-shaped nails used to be manufactured specifically for this purpose. Rather than put extra nails between the smallest slates in the four or five courses at the ridge that were the most exposed, it was usual to lay them in lime mortar, where they were known as tapping courses. Courses at the wallhead and at skews were also frequently bedded in mortar, and in fact, particularly in exposed locations, it was not unknown for this practice to be used over the whole roof.

A combination of small slates, inclement weather and the availability of sawn sarking influenced Scottish practice more or less in equal measure. What is clear is that this method of slating became well established during the 1700s and works very well. It was not thought out overnight: it evolved over a long period of time, and is a wonderful example of a traditional building skill, which looks superb and brings real individuality to many of Scotland's buildings.

FURTHER READING

Chapter 1: TIMBER

Alcock, N. W. *Cruck Construction: An Introduction and Catalogue*, London: Council for British Archaeology, 1981

Bruce, I. S. 'Timber frame construction in North East Scotland, a century of precedent 1830s–1930s', unpublished PhD thesis, The Robert Gordon University, Aberdeen, 2007

Brunskill, R. W. *Timber Building in Britain*, London: Gollancz, 1994 edition

Charles, F. W. B. *Medieval Cruck-building and its Derivatives*, London: Society for Medieval Archaeology, 1967

Cordingley, R. A. 'British historical roof-types and their members', *Transactions of the Ancient Monuments Society*, new series, 9 (1961), 73–129

Crone, A. and Mills, C. M. 'Dendrochronologically dated buildings from Scotland', *Vernacular Architecture*, 34 (2003), 84–9

Davies, I., Walker, B. and Pendlebury, J. *Timber Cladding in Scotland*, Edinburgh, 2002

Davey, N. *A History of Building Materials*, London, 1961

Dinwoodie, J. M. *Timber: Its Nature and Behaviour*, London, 2000 edition

Dixon, N. *The Crannogs of Scotland: An Underwater Archaeology*, Stroud: Tempus, 2004

Fawcett, R. (ed.) *Stirling Castle: The Restoration of the Great Hall*, York: Council for British Archaeology, 2001

Fawcett, R. *Scottish Medieval Churches: Architecture and Furnishings*, Stroud: Tempus, 2002

Hanke, T. 'The development of roof carpentry in South East Scotland until c.1650', Unpublished MPhil thesis, University of Edinburgh, 2005

Harris, R. *Discovering Timber-framed Buildings*, Princes Risborough, 1993 edition

Hay, G. D. 'Some aspects of timber construction in Scotland' in A. Fenton, G. Stell and B. Walker (eds.), *Building Construction in Scotland: Some Historical and Regional Aspects*, Dundee: SVBWG, 1976, pp. 28–38

Newland, K. 'Norwegian timber and the Scottish great house', *Architectural Heritage*, 18 (2007), 35–53

Ralston, I. *Celtic Fortifications*, Stroud, 2006

Reynolds, N. M. 'Dark Age timber halls and the background to excavations at Balbridie', *Scottish Archaeological Forum*, 10 (1978), 41–60

Smout, T. C. (ed.) *Scottish Woodland History*, Dalkeith: Scottish Cultural Press, 1997

Smout, T. C. (ed.) *People and Woods in Scotland: a History*, Edinburgh: Edinburgh University Press, 2003

Stell, G. 'Scottish Burgh Houses 1560–1707', in A. T. Simpson and S. Stevenson (eds), *Town Houses and Structures in Medieval Scotland*, Glasgow, 1980, pp. 1–31

Stell, G. 'Stone buildings with timber foundations: some unanswered questions', *Proceedings of the Society of Antiquaries of Scotland*, 114 (1984), 584–5

Stell, G. 'A note on medieval timber flooring and roofing' in A. Riches and G. Stell (eds), *Materials and Traditions in Scottish Building: Essays in Memory of Sonia Hackett*, Edinburgh: SVBWG, 1992, pp. 75–80

Yeomans, D. T. *The Trussed Roof: Its History and Development*, Aldershot: Scolar Press, 1992

Chapter 2: STONE

Adams, H. *Cassell's Building Construction*, London: Cassell & Co., 1906

Bremner, D. *The Industries of Scotland: Their Rise, Progress, and Present Condition*, Edinburgh: A. and C. Black, 1869

Howe, J. A. *The Geology of Building Stones*, London: Arnold, 1910

Hyslop, E., McMillan, A. and Maxwell, I. *Stone in Scotland*, Paris: UNESCO, 2006

Smith, P. *Rivington's Building Construction*, Vol. 3: *Materials*, Shaftesbury: Donhead, 2004 (first published Longman, 1904)

Whittow, J. B. *Geology and Scenery in Scotland*, Harmondsworth: Penguin, 1977

Chapter 3: BRICK

Bakewell, S. R. *Observations on Building and BrickMaking*, Manchester: Prentice and Cathrall, 1834

Brodribb, G. *Roman Brick and Tile*, Gloucester: Alan Sutton, 1987

Bremner, D. *The Industries of Scotland: Their Rise, Progress, and Present Condition*, Edinburgh: A. and C. Black, 1869

Curle, J. *A Roman Frontier Post and its People*, Glasgow: Maclehose, 1911

Douglas, G. and Oglethorpe, M. *Brick, Tile and Fireclay Industries in Scotland*, Edinburgh: RCAHMS, 1993

Hume, J. *The Industrial Archaeology of Scotland*, Vol. 2: *The Highlands and Islands*, London: Batsford, 1977

McKerrell, T. *Ayrshire Miners' Rows*, Ayrshire: Ayrshire Archaeological and Natural History Society, 1979

Reid, J. *The Scots Gard'ner*, Edinburgh: Mainstream, 1988 (first published Edinburgh, 1683)

Chapter 4: EARTH

Cornerstones Community Partnership. *Adobe Conservation: A Preservation Handbook* New Mexico: Sunstone Press, 2006

Houben, H. and Guillaud, H. *Earth Construction: A Comprehensive Guide*, London: Intermediate Technology Publications, 1994

Hurd, J. and Gourley, B. (eds) *Terra Britannica: A Celebration of Earthen Structures in Great Britain and Ireland*, London: James & James, c.2000

Keefe, L. *Earth Building: Methods and Materials, Repair and Conservation*. London: Taylor & Francis, 2005

Minke, G. *Earth Construction Handbook: The Building Material Earth in Modern Architecture*, Southampton: WIT Press, c.2000

Morton, T. *Earth Masonry: Design and Construction Guidelines*, Bracknell: IHS BRE Press, 2008

Walker, B. and McGregor, C. *The Hebridean Blackhouse: A Guide to Materials, Construction and Maintenance*, Edinburgh: Historic Scotland Technical Advice Note no. 5, 1996

Walker, B., McGregor, C. and Little, R. *Earth Structures and Construction in Scotland: A Guide to the Recognition and Conservation of Earth Technology in Scottish Buildings*, Edinburgh: Historic Scotland Technical Advice Note no. 6, 1996

Walker, P., Keable, R., Martin, J. and Maniatidis, V. *Rammed Earth: Design and Construction Guidelines*, Watford: BRE Press, 2005

Williams-Ellis, C. and Elizabeth Eastwick-Field, E. and J. *Building in Cob, Pise and Stabilized Earth*, London: Country Life, revised and enlarged edn,1947 (first published as *Cottage Building in Cob, Pisé, Chalk and Clay: A Renaissance*, 1919)

Chapter 5: CLAY

MacGibbon, D. and Ross, T. *The Castellated and Domestic Architecture of Scotland from the Twelfth to the Eighteenth Century*, Edinburgh: Douglas 1887

Mackenzie W. 'Clay castle building in Scotland', *Proceedings of the Society of Antiquaries of Scotland*, 68 (1934)

Walker B. *Scottish Turf Construction*, Edinburgh: Historic Scotland Technical Advice Note no. 30, 2006

Walker B. *Clay Building in North East Scotland*, Dundee: SVWG, 1977

Walker, B. McGregor, C. and Little, R. *Earth Structures and Construction in Scotland: A Guide to the Recognition and Conservation of Earth Technology in Scottish Buildings*, Edinburgh: Historic Scotland Technical Advice Note no. 6, 1996

Chapter 6: LIME

Allan, G. *Hydraulic Lime Mortar for Stone, Brick and Block Masonry*, Donhead, 2008

Fidler, J. 'A good rendering', in *Traditional Homes*, 1992

Gibbons, P. *Preparation and Use of Lime Mortar*, Edinburgh: Historic Scotland Technical Advice Note no. 1, 1995

Newsom, S. *External Lime Coatings on Traditional Buildings*, Edinburgh: Historic Scotland Technical Advice Note no. 15, 2001

Vicat, L. J. (trans. J. T. Smith) *A Practical and Scientific Treatise on Calcareous Mortars and Cements, Artificial and Natural*, London: John Weale, 1837

Wingate, M. *Small Scale Lime Burning*, London: ITP, 1985

Chapter 7: PLASTER

Apted, M. R. *The Painted Ceilings of Scotland: 1550–1650*, Edinburgh: HMSO, 1966

Bankart, G. P. *The Art of the Plasterer*, London: Batsford, 1908

Beard, G. *Decorative Plasterwork in Great Britain*, London: Phaidon, 1975

Butt, J. *The Industrial Archaeology of Scotland*, Newton Abbot: David & Charles, 1967

Gibbons, P., Newson, S. and Whitfield, E. *Care and Conservation of 17th Century Plasterwork in Scotland*, Edinburgh: Historic Scotland Technical Advice Note no. 26, 2004

Hume, J. R. *The Industrial Archaeology of Glasgow*, Glasgow: Blackie, 1974

Hume, J. R. *The Industrial Archaeology of Scotland*, vol. 2: *The Highlands and Islands*, London: Batsford, 1977

Hume, J. R. *The Industrial Archaeology of Scotland*, vol. 1: *The Lowlands and Borders*, London: Batsford, 1976

Mackay, S. *Behind the Façade: Four Centuries of Scottish Interiors*, Edinburgh: HMSO, 1995

Millar, W. *Plastering, Plain and Decorative*, London: Batsford, 1897

Naismith, R. J. *Buildings of the Scottish Countryside*, London: Victor Gollancz, 1985

Page, D. *Economic Geology or Geology in its Relations to the Arts and Manufactures*, Edinburgh & London: William Blackwood, 1874

Savage, P. *Lorimer and the Edinburgh Craft Designers*, Edinburgh: Paul Harris, 1980

Scottish Lime Centre Trust. *Charlestown Limeworks: Research and Conservation*, Research Report, Edinburgh: Historic Scotland, 2006

Chapter 8: PAINT

British Geological Survey. 'Mineral resource map for Clackmannanshire, Fife and Falkirk', OR/08/12, 2008

Brydall R. *Art in Scotland: Its Origin and Progress*, Edinburgh, 1899

General Register House, Edinburgh (ed. H. M. Paton). *Accounts of the Masters of Works for Building and Repairing Royal Palaces and Castles*, vol. 1, Edinburgh: HMSO, 1957

Gow, I. *The Scottish Interior*, Edinburgh: Edinburgh University Press, c.1992

Howard, H. *Pigments of English Medieval Wall Painting*, London: Archetype, 2003

Illustrated Sporting and Dramatic News, 4 May 1889

MacGibbon, D. and Ross, T. *The Castellated and Domestic Architecture of Scotland from the Twelfth to the Eighteenth Century*, vol. 2, Edinburgh: D. Douglas, 1887

Salzmann, L. F. *Building in England down to 1540*, Oxford: Oxford University Press, 1992

Stalker J. and Parker G. *A Treatise on Japanning and Varnishing*, London, 1960 (original edition 1688)

William, K. 'Restoring the magic to Law's Close, Kirkcaldy', *The Building Conservation Directory 2006*, Tisbury: Cathedral Communications, 2006

Chapter 9: IRON

Gloag, J. and Bridgewater D. *A History of Cast Iron in Architecture*, London: G. Allen and Unwin, 1948

Lister, R. *Decorative Wrought Ironwork in Great Britain*, Newton Abbot: David & Charles, 1970

Mitchell, D. S. and Laing A. C. *MacFarlane's Castings*, Harlaw: Heritage, 2004

184

Mitchell, D. S. *Walter MacFarlane & Co*, Edinburgh: Historic Scotland, 2009

Robertson, G. E. *Cast Iron Decoration – A World Survey*, London: Thames and Hudson, 1977

Ware, I. *A Complete Body of Architecture*, London: T. Osborne and J. Shipton, 1756

Chapter 10: LEAD

Barlow Bennett, S. *A Manual of Technical Plumbing*, London: Batsford 1910

Bremner, D. *The Industries of Scotland: Their Rise, Progress, and Present Condition*, Edinburgh: A. and C. Black, 1869

Parnell-Allen, J. *Practical Building Construction*, London: Crosby Lockwood and Son, 1893

Rivington's Notes on Building Construction, London: Longmans Green & Co, 1904

Weaver, L. *English Leadwork – Its Art and History*, London: Batsford, 1909

Chapter 11: GLASS

Barker, T. C. *The Glassmakers, Pilkington: 1826–1976*. London: Weidenfeld and Nicolson, 1977

Roche, N. *The Historical and Technical Development of the Sash and Case Window in Scotland*, Edinburgh: Historic Scotland, 2001

Turnbull, J. *The Scottish Glass Industry 1610–1750*. Edinburgh: Society of Antiquaries of Scotland, Monograph 18, 2001

Chapter 12: PANTILES

Board of Agriculture (London)/ Naismith, J. *General View of the Agriculture of Clydesdale*, London, 1806

Bo'ness and Blackness Import book 1680–81, National Archives of Scotland, MS NAS E72/5/9

Brunskill, R. W. *Illustrated Handbook of Vernacular Architecture*, London: Faber, 1970

Dunbar J. G. *The Historic Architecture of Scotland*, London, 1966

Fenton, A. and Walker, B. *Rural Architecture of Scotland*, Edinburgh: John Donald, 1981

Fidler, J. 'A clay roof over one's head', *Conservation Bulletin 33*, January 1998.

Lewis, S. *A Topographical Dictionary of Scotland*, London, 1846

MacGibbon, D. and Ross, T. *The Castellated and Domestic Architecture of Scotland, from The Twelfth to the Eighteenth Century*, Edinburgh: D. Douglas, 1887–92

McWilliam, C. *Buildings of Scotland: Lothian except Edinburgh*, Harmondsworth, 1978

Millan, B. 'The pantile experience', *Vernacular Building*, 27 (2003), 52–5

Naismith, R. J. *Buildings of the Scottish Countryside*, London: Victor Gollancz, 1985

Paton, H., Imrie, J. and Dunbar, J. G. (eds) *Accounts of the Masters of Works for Building and Repairing Royal Palaces and Castles*, Edinburgh: HMSO, 1957

Rolland, L. *Rossend Castle*, Edinburgh: Howie, 1977

Shaw, J. 'Dutch – and Scotch – pantiles: Some evidence from the seventeenth and early eighteenth centuries', *Vernacular Building*, 14 (1990), 26–9

Wilson, W. *Memoirs of the Life and Times of Daniel De Foe*, London, 1830

Chapter 13: THATCH

Fenton, A. and Walker, B. *The Rural Architecture of Scotland*, Edinburgh: John Donald, 1981

Holden, T. G. *The Archaeology of Scottish Thatch*, Edinburgh: Historic Scotland Technical Advice Note no. 13, 1998

Souness, J. 'Heather thatching in Scotland – further observations', *Vernacular Building*, 15 (1991), 3–26

Walker, B., McGregor, C. and Stark, G. *Thatches and Thatching Techniques: A Guide to Conserving Scottish Thatching Traditions*, Edinburgh: Historic Scotland Technical Advice Note no. 4, 1996

Chapter 14: SLATE

Adams, H. *Cassell's Building Construction*, London: Cassell & Co., 1906

Bennett, F. and Pinion, A. *Roof Slating and Tiling*, Shaftesbury: Donhead, 2000 (first published1948)

Bremner, D. *The Industries of Scotland: Their Rise, Progress, and Present Condition*, Edinburgh: A. and C. Black, 1869

Carmichael, J. 'An account of the principal marble, slate, sandstone and greenstone quarries in Scotland', *Prize Essays and Transactions of Highland and Agricultural Society*, 5 (1837), 398–416

Emerton, G., *Old Roofings*, London: NFRC, 1988

Emerton, G. *The Pattern of Scottish Roofing*, Edinburgh: Historic Scotland Research Report, 2000

Smith, P. *Rivington's Building Construction*, vol. 3: *Materials*, Shaftesbury: Donhead, 2004 (first published Longman, 1904)

Walsh, J. A. *Scottish Slate Quarries*, Edinburgh: Historic Scotland Technical Advice Note no. 21, 2000

ABOUT THE CONTRIBUTORS

Roz Artis-Young

Roz Artis-Young, Director of the Scottish Lime Centre Trust, has over 15 years' experience in building conservation through its advisory service Charlestown Consultants and its training arm Charlestown Workshops. Roz has a passion for encouraging the reinstatement of local vernacular finishes and colours, particularly limewashes which are authentic aesthetically and technically helping to preserve the regional identities of our built heritage. Roz has been a committee member of the Building Limes Forum and is currently on the management committees of both the National Heritage Training Group and Learn Direct and Build.

Stuart Eydmann

Stuart Eydmann was born and schooled in Fife. Since graduating from the Glasgow School of Art in European Architectural Heritage in 1975 he has worked as conservation officer in several local authorities. He has taught historical studies with the Open University and The Royal Scottish Academy of Music and Drama and has served as a university external examiner. Additionally Stuart researches and writes on urban conservation, architectural history and Scottish ethnography and is currently Conservation and Design Officer with West Lothian Council and Convener of the Institute of Historic Building Conservation in Scotland.

Tim Holden

Tim Holden is an archaeologist with long experience of working with Scottish traditional Building Materials and their remains. He has written widely on the subject, and on thatch in particular, being commissioned to author the Historic Scotland publication *The Archaeology of Scottish Thatch*. Tim is currently the Managing Director of Headland Archaeology.

Moses Jenkins

Moses Jenkins is currently a Senior Technical Officer at Historic Scotland. He has worked with the agency for four years having previously obtained degrees in history from the universities of Stirling and Glasgow. He is currently engaged in research projects in a number of areas related to built heritage including traditional brickwork in Scotland and energy efficiency measures in traditional buildings.

Roger Curtis

Roger Curtis spent time in the Royal Navy before completing a Master in Building Conservation in 2001. He joined the conservation contractor Cumming & Co., initially as a driver and latterly as project manager and safety officer. Conservation projects included Melgund Castle, Cathedral of The Isles, consolidation works at Drumin Castle and Saddell Abbey. He joined Historic Scotland in 2006 and is currently Head of Technical Research.

Neil Grieve

After almost twenty years working for local authorities in various capacities related to planning, Neil Grieve is now a lecturer in building conservation at the University of Dundee. He has produced a wide range of technical literature on subjects including *Scottish Slate: A Practitioners Guide* and *The Language of Urban Conservation*. Since 1991 he has been the Chief Executive of Tayside Building Preservation Trust and also a Director of the Scottish Urban Archaeological Trust.

Ingval Maxwell
Trained as an architect at the Duncan of Jordanstone College of Art, Dundee, Ingval Maxwell joined Historic Scotland's predecessor, the Ancient Monuments Branch of the Ministry of Public Buildings and Works, in 1969. He fulfilled the roles of Area Architect, Principal Architect, Assistant Director and, finally, Director of Technical Conservation, Research and Education (TCRE) before retiring from Historic Scotland in 2008.

Chris McGregor
Chris McGregor has a great interest in vernacular building, kindled by Dr Bruce Walker at Dundee University, who was his tutor at the School of Architecture.
Having worked in private practice for several years he took an opportunity which arose in 1986 to work on Stirling Castle as Project Architect with Historic Scotland. He has remained with Historic Scotland since then, working on a large range of projects from Sunnybrae Cottage in Pitlochry to Stanley Mills, a substantial Cotton Mill just north of Perth. He has co-authored a number of publications on the traditional vernacular building of Scotland and is an active member of the Scottish Vernacular Buildings Working Group. He now works in the Technical Conservation Group continuing to look at traditional buildings.

David Mitchell
Prior to joining Historic Scotland in 2002, David Mitchell was Managing Director of a private firm of Industrial Heritage Consultants and Contractors based in Glasgow for ten years. His work in the Technical Conservation, Research and Education (TCRE) Group centred around technical and scientific research and traditional building skills; he was appointed as Director of the Technical Conservation Group in 2008. He has an academic background in earth sciences and is currently working towards a PhD in Architecture. He is a Trustee of The Scottish Ironwork Foundation and the Scottish Industrial Heritage Society.

Robin Murdoch
After a long career as an engineer, Robin Murdoch is now Director of Harlaw Heritage, a small company involved in archaeology and associated historical research. He is an expert on historic glass and is frequently called upon to discuss glass finds from archaeological digs throughout Scotland.

Michael Pearce
Michael Pearce has researched historic interiors with Historic Scotland since the late 1990s. Recent projects include contributing to the redecoration of the dome at General Register House, the nave of St Giles in Edinburgh, and the re-instatement of the royal apartments at Stirling Castle. Currently, Michael is investigating burgh records to shed light on the work of Scottish renaissance decorative painters. Michael has an MPhil in Scottish Architectural History, and came to Scotland after working in architectural conservation in England.

Geoffrey Stell
Geoffrey Stell was trained as an historian at Leeds and Glasgow universities where he specialised in French and Scottish medieval history. In 1969 he joined the Royal Commission on the Ancient and Historical Monuments of Scotland (RCAHMS) as an Investigator of Historic Buildings, and from 1991 until his retirement in 2004 he was Head of Architecture, responsible for all RCAHMS architectural survey and recording work. In addition to his association with the Department of History at the University of Stirling, he is also Visiting Lecturer in the Department of Architecture at Edinburgh College of Art.

Gordon Urquhart

Gordon Urquhart was born in New York City. He first returned to his father's native Scotland in 1979 to study Scottish History as an exchange student at King's College, Aberdeen. Between taking a BA in History at Hamilton College, New York, and an MSc in Architectural Conservation at Columbia University, he lived and worked in Edinburgh's New Town. For fifteen years he was Assistant Director with the Glasgow West Conservation Trust, during which time he co-wrote *The West End Conservation Manual*. Gordon has written and lectured widely on traditional building technology, architectural conservation and local history. In 2000 he published *Along Great Western Road*, the first comprehensive history of the West End of Glasgow.

Bruce Walker

Bruce Walker has worked previously as both a lecturer at Dundee University and an architect with Historic Scotland. He has a long and well developed interest in traditional buildings in Scotland and was a founding member of the Scottish Vernacular Buildings Working Group. He is the author of numerous books and articles on Scotland's traditional buildings including co-writing *The Rural Architecture of Scotland* with Alexander Fenton.

Sam Sills: *photographer*

Sam Sills is a highly experienced photographer having worked on a wide range of projects for various commercial clients. Having built up a reputation for bringing vibrancy to her subjects she was chosen to take the images used in this volume to help present traditional building materials in a fresh and dynamic way.

INDEX

Note: entries in italic type refer to illustrations.

192

195